Culture of Salmonid Fishes

Editor

Robert R. Stickney
Professor
School of Fisheries
University of Washington
Seattle, Washington

CRC Press
Boca Raton Ann Arbor Boston London

Library of Congress Cataloging-in-Publication Data

Culture of salmonid fishes / editor, Robert R. Stickney.
p. cm.
Includes bibliographical references (p.) and index.
ISBN 0-8493-5310-6
1. Salmon. 2. Trout. 3. Salmonidae. 4. Fish-culture.
I. Stickney, Robert R.
SH167.S17C85 1991
639.3′755--dc20 91-4104
CIP

This book represents information obtained from authentic and highly regarded sources. Reprinted material is quoted with permission, and sources are indicated. A wide variety of references are listed. Every reasonable effort has been made to give reliable data and information, but the author and the publisher cannot assume responsibility for the validity of all materials or for the consequences of their use.

Direct all inquiries to CRC Press, Inc., 2000 Corporate Blvd., N.W., Boca Raton, Florida 33431.

International Standard Book Number 0-8493-5310-6

Library of Congress Card Number 91-4104

Printed in the United States of America 1 2 3 4 5 6 7 8 9 0

PREFACE

Aquaculture is the rearing of aquatic organisms under controlled or semi-controlled conditions.[1] That definition carries no implication that aquacultured products are reared with a profit in mind or that they are strictly reared for human consumption, though both objectives (often in combination) commonly exist. With respect to finfish aquaculture, particularly in North America, the objective is often to produce fish for stocking into sport and/or commercial fisheries.

Various species of trout and salmon, members of the family Salmonidae, are cultured in temperate and sub-Arctic climates around the world. In North America, salmonid fishes are cultured in nearly all 50 of the United States as well as in the provinces of Canada. Six species of salmon are produced, along with various species of trout. Spawning, egg incubation, and early rearing strategies are virtually the same whether the fish are to be released into the natural environment or maintained in captivity for extended periods. We have attempted to avoid redundancy in the chapters which follow by presenting early rearing methods only once for any given species. In general, chapters dealing with production of fish for release cover the early life history stages, as for example, with regard to Pacific salmon. That information is not repeated in subsequent chapters that deal with Pacific salmon.

Species discussed below were selected on the basis of their importance to the aquaculture community in the broad sense. Several additional species have been, or are being, cultured to various extents, but the bulk of the production involves the species or species groups discussed in the various chapters that follow.

The same principles of fish culture are generally applicable to all species. Culture techniques vary with respect to details, of course, but the needs for adequate culture systems, suitable water quality, properly formulated and manufactured feeds, disease control, and so forth are universal. Aquaculturists do tend — perhaps by convention or the temperature difference that is readily detected by the human senses — to make a distinction between warmwater and coldwater species. While there is really no logical reason for this separation, practitioners typically consider themselves to be specialists in the culture of warmwater fishes or coldwater fishes, rarely both. There is also what is arguably an illogical distinction by many between those who work with marine species as opposed to those who limit their activities to freshwater. The arbitrary nature of the latter distinction is obviated when one considers anadromous species (such as salmon) or catadromous species (such as eels).

In a companion volume, *Culture of Nonsalmonid Freshwater Fishes,*[1] we bowed to convention in two ways: by considering warmwater species and the so-called midrange species (those with a temperature optimum between those of coldwater and warmwater fishes), and by considering only freshwater species. In this volume, we concentrate on the salmonids, which represent the primary coldwater fishes of culture interest in the world today.

In recognition of the growing worldwide concern that aquaculture development has the potential, which in some cases has actually been realized, to negatively impact the natural environment, a chapter on controversies surrounding salmonid culture has been included in this book. That approach is a departure from the initial volume,[1] as is the first chapter, which provides a summary of life history information of the species covered in the book.

Robert R. Stickney
Seattle, Washington

THE EDITOR

Robert R. Stickney, Ph.D., is a professor in the School of Fisheries at the University of Washington, Seattle, Washington. The School of Fisheries is the oldest academic fisheries program in the United States and has a faculty in excess of 30 and a student enrollment of about 200, the majority of whom are graduate students.

Dr. Stickney took his B.S. and M.A. in Zoology at the University of Nebraska (1967) and the University of Missouri (1968). He received his Ph.D. in Oceanography from Florida State University in 1971. He was on the staff at the Skidaway Institute of Oceanography in Savannah, Georgia from 1971 to 1975, then joined the faculty of the Department of Wildlife and Fisheries Sciences at Texas A&M University, College Station, Texas. In 1984, Dr. Stickney accepted the position of Director of the Fisheries Research Laboratory at Southern Illinois University in Carbondale, Illinois. He accepted his current position in the summer of 1985.

Dr. Stickney's research has involved studies on aquatic organisms ranging from phytoplankton to marine mammals, but he has concentrated his work in the area of fish culture. He and his graduate students have conducted extensive studies on channel catfish, tilapia, and more recently, salmon, trout, and Pacific halibut. Dr. Stickney has been primarily interested in fish nutrition and the environmental requirements of fish under culture conditions.

Over 100 scientific papers have been authored or co-authored by Dr. Stickney. He has also written or been involved in editing eight books. He is a certified fishery biologist and member of the American Fisheries Society and has served as President of the Fish Culture Section and the Education Section of the Society. He is a member of the World Aquaculture Society and has served on its Board of Directors since 1987. He was elected to the office of President-elect in 1990 and assumed the presidency of the World Aquaculture Society in 1991. Dr. Stickney is also a member of the American Association for the Advancement of Science, the American Society of Limnology and Oceanography, the American Institute of Nutrition, and is a Fellow of the American Institute of Fishery Research Biologists.

Dr. Stickney serves as Chairman of the Board of Institutional Advisors for the Fish Stock Assessment Collaborative Research Support Program of the Agency for International Development, and on the Board of Directors of the Western Region Aquaculture Consortium (past Board Chairman). He has been actively involved in international development with respect to aquaculture and fisheries.

CONTRIBUTORS

Ernest L. Brannon
Aquaculture Institute
University of Idaho
Moscow, Idaho

David Fluharty
School of Marine Affairs
University of Washington
Seattle, Washington

Arni Isaksson
Institute of Freshwater Fisheries
Reykjavik, Iceland

Conrad Mahnken
National Marine Fisheries Service
Manchester, Washington

John D. McIntyre
Intermountain Research Station
USDA-Forest Service
Boise, Idaho

Robert R. Stickney
School of Fisheries
University of Washington
Seattle, Washington

TABLE OF CONTENTS

Chapter 1

SALMONID LIFE HISTORIES

Robert R. Stickney

TABLE OF CONTENTS

I. INTRODUCTION

Traditionally, mention of the word trout has often conjured up the image of a person in waders, standing in a clear mountain stream and casting a dry fly into a pool. On the other hand, the word salmon may elicit a mental picture of bears pulling bright red fish out of the water during the spawning season. These are just two possibilities among many, but in North America, at least, most people are familiar with both groups of fishes, whether they are personally involved in fishing or just see trout and salmon in their local supermarkets.

Trout have been produced in hatcheries for well over a century. Various native species have been widely introduced around the country, and a European cousin was brought in many decades ago to augment existing sport fisheries. In many states, sportfish populations of trout would not be sustained without augmentation stocking. Some populations are maintained solely by stocking as there is little or no natural recruitment.

Commercial trout culture began in earnest within the past few decades, and it was not until the 1970s that the industry began to mature. Today, the production of trout for direct sale in restaurants and supermarkets is centered in the Hagerman Valley of Idaho where large underground streams outcrop along the Snake River Canyon and pour thousands of cubic meters per second of almost ideally suited water into the Snake River. A significant portion of that water is now intercepted and run through raceways on aquaculture facilities which produce the bulk of all commercially produced trout in the nation.

Up until a few years ago, the only access most people in the United States had to salmon was from a can. Today, fresh salmon fillets and steaks are available in most major cities, typically on a year round basis. Fish from the wild contribute to the fresh fish market during the portion of the year when they are available to commercial fishermen, while cultured fish are responsible for expanding the market nationwide and making the product available through the year. Some cultured salmon are produced in the United States, but a large percentage comes from Norway and, increasingly, Canada, Chile, Scotland, and a few other countries are also culturing significant amounts of salmon.

In this book, we examine the production of fishes within the family Salmonidae, which includes the various species of trout and salmon. Some of the material deals with the extremely important subject of producing fingerlings for release by government agencies into the natural environment for enhancement of sport and, in some cases, commercial fisheries. The subject of ocean ranching, wherein private enterprise produces fish, releases them into the marine environment, and harvests their own stock when it returns, is also covered in these pages. Trout and salmon are also being produced entirely in captivity, primarily in raceways, marine net-pens, and freshwater cages, and those techniques are fully described.

This chapter describes the basic life histories of the important species of trout and salmon. While many of the species share similar environmental requirements, habitats, and behavioral patterns, there are significant distinctions among them, some of which impact culture strategies. The culture of many of the species outlined in this chapter is detailed later in the book.

II. TAXONOMY

Trout and salmon are members of the family Salmonidae. The family is characterized by torpedo-shaped fishes with soft fin rays and adipose fins. Salmonids produce large eggs (several mm in diameter), the sexes are separate, and spawning generally occurs in the gravel of streams or, less commonly, along the shorelines of lakes or in the brackish water areas of estuaries.

TABLE 1
Scientific and Common Names of Selected Salmonid Fishes[2,3]

Scientific name	Common name
Oncorhynchus gorbuscha	pink salmon
Oncorhynchus keta	chum salmon
Oncorhynchus kisutch	coho salmon
Oncorhynchus mykiss (formerly *Salmo gairdneri*)	rainbow trout
Oncorhynchus nerka	sockeye salmon
Oncorhynchus tshawytscha	chinook salmon
Salmo clarki	cutthroat trout
Salmo salar	Atlantic salmon
Salmo trutta	brown trout
Salvelinus fontinalis	brook trout
Salvelinus namaycush	lake trout

The primary genera of interest to culturists are *Salmo* and *Oncorhynchus,* and to a lesser extent, *Salvelinus.* It has generally been true in the past that trout were placed in the genera *Salmo* and *Salvelinus,* while salmon were in the genus *Oncorhynchus.* The exception was the Atlantic salmon, *Salmo salar,* which, because of the genus to which it has been assigned, has been considered a trout by many scientists.

Added confusion was created in 1989 when the rainbow trout, *Salmo gairdneri,* was reclassified and placed in the genus *Oncorhynchus.*[2] Today, rainbow trout are recognized as *O. mykiss,* which may technically make them a species of salmon rather than trout. The fact that the steelhead trout is recognized as a sea-run strain of the rainbow (both being *O. mykiss)* provides credibility for the argument, though cutthroat trout (*Salmo clarki*) also have strains which migrate to saltwater for portions of their lives.

To avoid clouding the issue, throughout this book we utilize the American Fisheries Society taxonomy for both scientific and common names,[3] with the exception that we have adopted the new taxonomy for rainbow trout[2] because it has now been accepted by the American Fisheries Society.

III. LIFE HISTORY INFORMATION

Included in the family Salmonidae are various species commonly known as cisco, trout, salmon, whitefish, and grayling. Those which are of current interest to fish culturists are included in Table 1. Listed in that table are such species as chum and pink salmon which have not received anything comparable to the attention that has been focussed on coho salmon, chinook salmon, Atlantic salmon, and rainbow trout. There are hatcheries, particularly those associated with ocean ranching operations in Alaska, which are producing chum and pink salmon. Sockeye salmon were studied for a period in the past and found difficult to rear, but recent attempts to revisit that species have resulted in the development of what appears to be dependable culture practices.[4] Interest in the culture of Arctic char (*Salvelinus alpinus*) is developing, and that species could become a commercial aquaculture species at high latitudes where temperatures do not support rapid growth of other salmonid species.

FIGURE 1. Pink salmon (*Oncorhynchus gorbuscha*) immature form. (Computer-generated drawing by Kenneth Adkins, School of Fisheries, University of Washington, Seattle.)

A. PINK SALMON (*ONCORHYNCHUS GORBUSCHA*)

Of the five species of Pacific salmon which are native to North America, the pink salmon is one of the smallest. It reaches up to 76 cm in length and weighs from 1.4 to 2.3 kg at maturity. Spawning populations are distributed in rivers and small streams most commonly from northern California to Alaska. They can also be found in the Aleutian, Commander, and Kuril Islands of the North Pacific Ocean.[5]

The physical characteristics of pink salmon (Figure 1) have been described as follows:[5]

- Fins — dorsal fin with 10 to 15 rays
  caudal fin slightly forked
  anal fin with 13 to 17 rays
  pectoral fins each with about 15 rays
  pelvic fins each with about 10 rays
- Gill rakers — 26 to 34 on the first arch
- Scales — 150 to 205 on the lateral line

In general, pink salmon are metallic blue on their dorsal surface and silvery on the sides. There are numerous black, oval spots on the upper sides of the body, the back, and the caudal fin. Characteristic of species within the genus, mature fish take on different colors from the immature form. Mature males are red to yellow on the sides of the body with blotches of brown; they are dark along the back. Females are olive green on the sides of the body with dusky stripes.[5]

As is true of the other species of Pacific salmon, pink salmon are anadromous. They spawn in freshwater, spend a portion of their life in that environment, go through a process called smoltification which physiologically prepares them for entry into the marine environment, then migrate to the sea for a period of time. They return to their natal stream to spawn upon reaching maturity. A summary of the life history patterns of the five species of Pacific salmon found in North America is presented in Table 2.

Pink salmon spawning has been described by Bailey.[6] Spawning occurs during late summer or early fall, depending upon the location of the spawning stream. As is generally true for all species of Pacific salmon, spawning occurs later in the year at the lower latitudes and earlier at high latitudes. Spawning typically occurs in short streams, though the species has been known to spawn at least 330 km from the mouth of certain streams in British Columbia and California, and twice that far upstream in Asia. The fish may also spawn in

TABLE 2
Life Cycle Patterns of Five Species of North American Pacific Salmon in Alaska Waters[8]

Species	Freshwater habitat type	Time in freshwater after fry emergence	Time at sea (yr)
Pink	short streams and lakes	usually less than 1 day	1
Chum	short and long streams	less than 1 month	2—4
Coho	short streams and lakes	12—24 months	1—3
Sockeye	short streams and lakes	12—36 months	1—4
Chinook	large rivers	3—12 months	1—4

the lower reaches of short streams or even in intertidal areas where the eggs may be alternately exposed to fresh and brackish water during incubation.

Female pink salmon, again typical of the other Pacific salmon, construct nests, called redds, in the spawning gravel. The redds may be from 10 to 25 cm deep and usually occur in riffles.[6] Females generally release from about 1,500 to 1,900 eggs[5] which are fertilized as they fall into the redd and become deposited in the interstices that exist in the gravel.

Incubation requires up to several months, depending upon temperature, with hatching occurring in the spring of the year following spawning. Normal development will not occur at constant temperatures below about 5°C, though exposure to that temperature for a month before exposure to 2°C apparently results in normal larval development.[7] During the spawning season eggs may be destroyed as females construct redds in areas which were previously utilized by other fish. Other losses can occur because of low dissolved oxygen levels in the gravel, eggs becoming dislodged by floods, mortality caused by freezing temperatures, and predation. The percentage of eggs that survive until emergence of the fry is often less than 25%.[6]

Upon hatching, larval pink salmon remain in the gravel for a period, usually a few weeks, during which they obtain nutrition from their large yolk sacs. Pink salmon fry emerge from the gravel at night and migrate directly to the sea (Table 2). The young fish form schools near the surface once they reach the estuary and will then migrate along the shore. They are carried out to sea by currents within a few days to several weeks.[6]

Pink salmon are predaceous sight feeders. They initially feed on planktonic organisms but adjust their food habits to squid and other fishes as they reach sizes where they can capture and consume those types of prey.[6] The fish prefer temperatures between 7 and 15°C in the Gulf of Alaska.[5] They return to spawn in their second year of life.[8] As is true of all Pacific salmon species, the adults die soon after spawning.

B. CHUM SALMON (*ONCORHYNCHUS KETA*)

Chum, also known as dog salmon, occur from southern California to Alaska. They can be found in the Aleutian, Commander, and Kuril Islands, as well as in the Siberian Arctic region, and are distributed southward in Asia to Japan.[5] The chum salmon is the most widely distributed of the Pacific salmon species and is second in abundance.[8] Distinguishing characteristics are as follows:[5]

- Fins — dorsal fin with 10 to 13 rays
  caudal fin slightly forked
  anal fin with 13 to 17 rays
  pectoral fins each with about 16 rays
  pelvic fins each with about 10 rays

FIGURE 2. Chum salmon (*Oncorhynchus keta*) immature form. (Computer-generated drawing by Kenneth Adkins, School of Fisheries, University of Washington, Seattle.)

- Gill rakers — 18 to 26 on the first arch
- Scales — 126 to 131 on the lateral line

Chum salmon are metallic blue in color on their dorsal surfaces with occasional black speckling but no distinct black spots. The tips of the pectoral, anal, and caudal fins are dark at their tips (Figure 2). When they mature in freshwater, they develop reddish or dark streaks or bars. The species has been known to reach lengths slightly in excess of 100 cm and may weigh up to 15 kg.[5]

Spawning occurs during late summer and fall in streams of various lengths[5,8] (Table 2). Most fish spawn from June to January, with northern populations spawning by the end of August or early in September.[9] Southern populations spawn later.[10,11] Spawning sites are known from the United States, Canada, Japan, Korea, and the U.S.S.R.[9] Adults generally select gravelly riffles in habitats ranging from tidal flats associated with small streams to springs in the headwaters of larger stream systems. The Yukon River run is the longest, with spawners being found over 2,500 km upstream from the river mouth.[8]

Fry emerge from the gravel in the spring and migrate directly to sea, with the trip requiring up to a month in streams where the young fish have to negotiate long distances (Table 2). Those which have short distances to traverse between the spawning grounds and the sea do not feed until they reach the ocean. During their first summer at sea they consume small invertebrates. The food habits of chum salmon change to larger invertebrates and fish as they grow and are able to catch and consume the larger prey species. The fish spend 2 to 4 years at sea and return to spawn at 4 to 5 years of age.[8]

C. COHO SALMON (*ONCORHYNCHUS KISUTCH*)

The coho salmon (also known as the silver salmon) is a popular sport fish in the Pacific Northwest and is one of the salmon species that has been introduced to the Great Lakes where a large sport fishery has developed. The first coho salmon eggs, over one million in number, were introduced into Lake Michigan by the Michigan Department of Natural Resources in 1966. Introduction of the fish to the other Great Lakes was conducted in both the United States and Canada in 1968 and 1969. By 1983, the total number of fish planted in the Great Lakes had reached 83.9 million.[12]

Natural coho salmon populations occur most commonly from Monterey, California north to Point Hope, Alaska. The species has been reported as far south as Chamalu Bay, Mexico.[13] Coho are found throughout the Aleutian Islands, and from the Anadyr River in the U.S.S.R., southward to Korea and Hokkaido, Japan.[14] The fish are most abundant between Oregon

FIGURE 3. Coho salmon (*Oncorhynchus kisutch*) immature form. (Computer-generated drawing by Kenneth Adkins, School of Fisheries, University of Washington, Seattle.)

and southeast Alaska[5] and were rare in the Sacramento River system of California until the California Department of Fish and Game stocked them in 1956 to 1958.[15] They have the following characteristics:[5]

- Fins — dorsal fin with 9 to 13 rays
  caudal fin slightly indented
  anal fin with 13 to 16 rays
  pectoral fins each with about 15 rays
  pelvic fins each with about 11 rays
- Gill rakers — 19 to 25 on the first arch
- Scales — 118 to 147 on the lateral line

Immature fish are metallic blue on the dorsal surface and silvery on the sides, ventral surface, and caudal peduncle. There are irregular black spots on the back and upper lobe of the caudal fin (Figure 3). Maturing males in freshwater are bright red on the sides, with bright green coloration on the back and head. The mature females are less strongly colored.[5]

Entry of coho salmon into freshwater begins in July throughout most of the range of the species, with upstream runs occurring from August to February. Typically, the adults spend 30 to 60 days in freshwater, with North American fish demonstrating peak spawning activity between September and January, though some fish will spawn as late as March. Spawning usually occurs in small streams, but the species has also been known to spawn in large rivers, particularly within 240 km of their mouths (Table 2). Spawning habitat typically includes current velocities as fast as 0.3 to 0.5 m/sec as compared with 0.1 m/sec for sockeye salmon.[14]

Coho salmon have been captured at sea as far as 1,930 km from their point of origin, though they tend to remain within a few hundred kilometers of the coast. They are usually found within 10 m of the sea surface except when the surface water is warm. Some fish are nearshore residents throughout the period of their lives spent in the marine environment. There are groups of coho salmon, for example, which spend the entire seawater period in the waters of Puget Sound, Washington, while others migrate out into the Pacific Ocean after variable periods in the Sound.[14]

In Kamchatka, U.S.S.R., spawning occurs at temperatures between 0.8 and 7.7°C,[16] while on the west coast of the U.S., spawning has been reported over a temperature range of 4.4 to 9.4°C.[17] When held at constant incubation temperatures, coho embryos will withstand temperatures between 1.3 and 12.4°C.[18] Small changes in temperature can lead to relatively rapid changes in development rate. For example, incubation requires 38 days at 11°C, 48 days at 9°C, and 86 to 101 days at 4.5°C.[14,19]

Fecundity varies as a function of the size of the spawning female, the geographic area where the fish are found, and the individual year.[14] The general range is from 1,440 to 5,700 eggs for females ranging from 44 to 72 cm long in Washington.[20] Fecundity in coho salmon can be calculated from the following formula:[19]

$$\text{Number of eggs} = 0.01153 \times \text{fork length}^{2.9403}$$

The eggs are demersal, red in color, and from 4.5 to 6.0 mm in diameter.[20] A female may deposit eggs in three to four different redds which she constructs by lying on her side and beating the gravel with her tail.[14]

The newly hatched fish, called alevins, begin emerging from the gravel 2 to 3 weeks following hatching and may continue to emerge for an additional several weeks.[21] They are photonegative at first, but become photopositive with time. Emergence occurs from March to July. Fry live in shallow gravel areas, at first forming schools, which later disperse. The fry are most attracted to water temperatures between 10 and 15°C and dissolved oxygen concentrations near saturation; they are typically found in riffles that have few sediments.[17,22,23]

As the fish grow to fingerling size in freshwater, they are known as parr. Parr are characterized by the development of 8 to 12 distinct vertical dark bars along their sides. The bars are narrower than the interspaces between them.[24]

Some coho salmon migrate to sea during their first year in freshwater, but migration during the second year is more common (Table 2). Two years in freshwater is typical for fish produced in the Yukon River drainage, for example.[20] If the fish are grown in warmer than normal water, which is sometimes possible under hatchery conditions, growth to the size of smolting can be accelerated. A strain of fish at the University of Washington has been adapted to smolt at 6 months instead of the normal 18 months by rearing the fingerlings in water of approximately 9 to 13°C.[25] Those fish returned from sea after 18 months as compared with the normal 2 years at sea.

The movement of fish from freshwater to the marine environment, known as outmigration, occurs primarily at night.[26] Smoltification may occur earlier than normal when freshwater temperatures are unusually high.[27] This can lead to outmigration of young coho salmon into marine conditions that are unfavorable, e.g., limited food resources.

D. SOCKEYE SALMON (*ONCORHYNCHUS NERKA*)

The sockeye, or red salmon, is found in commercial quantities along the North American coast from the Columbia River northward to Bristol Bay, Alaska. The species occurs throughout the Aleutian, Kuril, and Commander islands in the North Pacific, and there is a large population found around the Kamchatka peninsula and in the northern Sea of Okhotsk. Sockeye salmon have been observed in Oregon waters of North America and in northern Hokkaido, Japan, but those sightings are relatively unusual. Characteristics of the species, which is depicted in Figure 4, are as follows:[5]

- Fins — dorsal fin with 11 to 16 rays
  caudal fin moderately forked
  anal fin with 13 to 18 rays
  pectoral fins each with about 16 rays
  pelvic fins each with about 11 rays

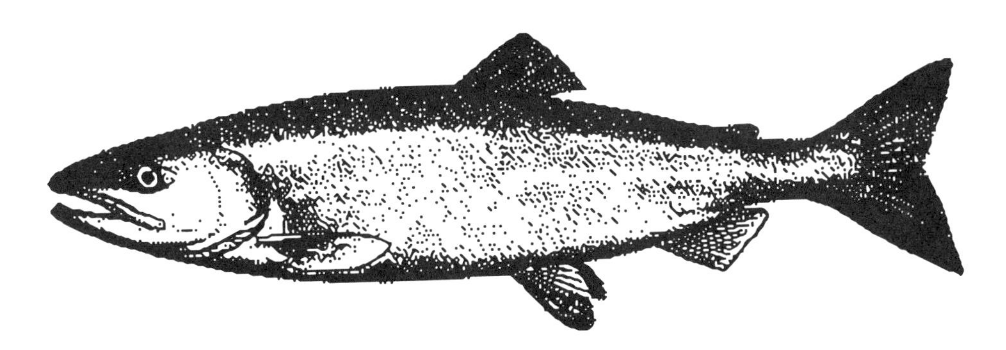

FIGURE 4. Sockeye salmon (*Oncorhynchus nerka*) immature form. (Computer-generated drawing by Kenneth Adkins, School of Fisheries, University of Washington, Seattle).

- Gill rakers — 28 to 40 on the first arch
- Scales — 125 to 145 on the lateral line

Sockeye reach lengths up to 84 cm[5] and may weigh up to 6 to 7 kg.[28] Fish from different river systems have sizes characteristic of their native stocks.[5] Spawners range in age from 2 to 7 years, with most being of ages intermediate to those extremes.[29]

The species is unusual in that it spawns in both streams and lakes (Table 2). Spawning takes place in the late summer or autumn in river inlets and outlets associated with lakes and in the lakes themselves. The females construct redds that are 25 to 40 cm deep and may occur at water depths as great as nearly 30 m. In Alaska, where the largest North American concentrations of sockeye are found, spawning occurs when the water temperature ranges from about 5 to 10°C.[29]

Fecundity ranges from about 2,200 to over 4,300 eggs per female, with the average being 3,720. Incubation can require as little as 50 days and as long as 5 months, depending upon temperature.[5] The lower temperature threshold for normal development of sockeye embryos is reportedly about 5 to 6°C, while the upper temperature limit for egg development is 13 to 14°C. If sockeye eggs are maintained at 6°C for 96 hours following fertilization, they can withstand temperatures as low as 2°C for the remainder of the developmental period.[30]

Sockeye fry emerge from the gravel in the winter and form schools. Those that emerge in streams migrate to lakes, while those spawned in lakes remain in those bodies of water. While in freshwater they are subject to predation by char, sculpins, trout, and birds.[29] In Lake Washington, predation from squawfish has also been reported.[31]

The rivers associated with Bristol Bay, Alaska often have large lakes associated with them, some of which have been heavily studied by salmonid biologists. Lake Iliamna and the Wood River lake system, which features a series of large lakes connected to one another by short stretches of river, are two systems that have been studied since the mid-1940s by scientists associated with the Fisheries Research Institute of the University of Washington.[32] Annual estimates of Bristol Bay sockeye runs are produced by biologists associated with the University of Washington and the Alaska Department of Fish and Game who study the Alaska lake systems.

Young sockeye spend from 1 to 3 years in freshwater and an additional 1 to 4 years at sea as shown in Table 2, though some authors give a range of 1 to 4 years in freshwater and 1 to 3 years at sea.[29] Outmigration of smolts usually occurs at night. When the fish reach the sea they feed first on planktonic crustaceans and eventually convert to shrimp, squid, and small fish as the salmon attain sufficiently large sizes to accept those prey items.

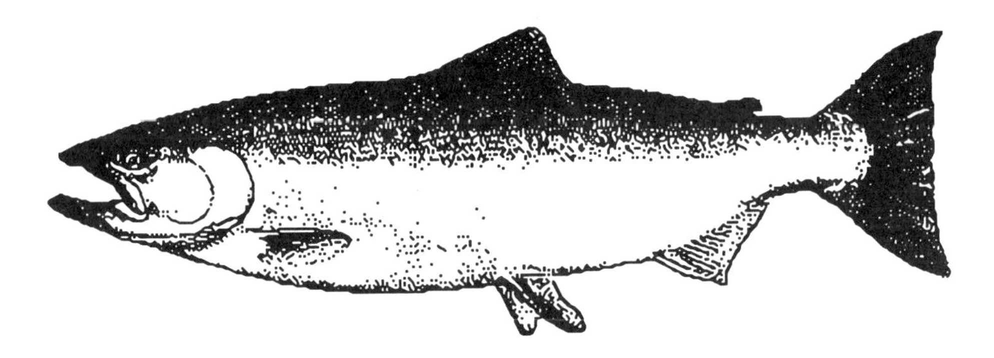

FIGURE 5. Chinook salmon (*Oncorhynchus tshawytscha*) immature form. (Computer-generated drawing by Kenneth Adkins, School of Fisheries, University of Washington, Seattle.)

E. CHINOOK SALMON (*ONCORHYNCHUS TSHAWYTSCHA*)

The chinook, or king salmon (Figure 5), is the largest of the Pacific Ocean species, and is highly prized as a sportfish. The famous Kenai River salmon run, for example, is fished on by thousands of anglers in search of trophy chinooks that may exceed 30 kg in weight. The record for the species is over 57 kg, though the common range of chinooks seen in both the sport and commercial catches is generally from 6 to 23 kg.[33] The fish has the following characteristics from which it can be identified:[5]

- Fins — dorsal fin with 10 to 14 rays
  caudal fin moderately forked
  anal fin with 13 to 19 rays
  pectoral fins each with about 14 rays
  pelvic fins each with about 10 rays
- Gill rakers — 18 to 30 on the first arch
- Scales — 130 to 165 on the lateral line

Chinook are greenish blue to black on the dorsal surface. They have irregular black spots on the dorsal fin, both lobes of the caudal fin, back, and upper sides. There is also black along the bases of the teeth;[5] thus, the fish is also called the blackmouth.

The original range of the chinook salmon was from the Ventura River, California to Point Hope, Alaska in the northwestern Pacific Ocean. In the northeastern Pacific, the species has been found from the Anadyr River in the U.S.S.R. to Hokkaido, Japan.[34,35] Spawning populations now occur from the Sacramento-San Joaquin River system in California north to Point Hope, Alaska.[36] Chinook spawn in some 640 streams along the Pacific coast, with major populations occurring in the Sacramento, Columbia, Copper, Nushagak, Kuskokwim, and Yukon Rivers of the U.S. and some 14 rivers in Canada.[37]

Chinook salmon, like coho, have been planted in the Great Lakes and contribute significantly to the sport fishery that has developed in those bodies of water. Chinook plantings began in lakes Michigan and Superior in 1967. By 1983, 108 million fish had been stocked. The fingerlings are maintained for 6 months in hatcheries before release. The stock utilized in the Great Lakes matures in 3 to 4 years.[12]

There are four recognized races of chinook salmon: spring, summer, fall, and winter. There is a great diversity which exists in races of chinook with respect to time of entry of the adults into river systems, the timing of their spawning activities, the distance that they travel from sea to the spawning grounds, the length of time that they spend in both freshwater and the marine environment, and their average age and size at maturity (Table 2).[33] The

Columbia River is an example of this population diversity in action. Three races are recognized in the Columbia River on the basis of time of entry into the river, time of spawning and age, and time of smolt outmigration. Spring chinook enter from February through May, summer chinook from June through August, and fall chinook from mid-August through October.[34,38]

As one moves from the south to the north, the following tendencies are found:

- The number of races decreases from four in the Sacramento River to three in the Columbia, Fraser, and Nanaimo Rivers, to two in southeastern Alaska, and one in northern Alaska[33]
- Spawning occurs earlier in the year, with peaks coming in late August through October in the south and July through August in the north[33]
- The length of freshwater residence increases from a few weeks to a year in the southern portion of the range to an average of 1 to 2 years in the north[39]
- Average age at spawning increases from 3 to 6 years in California to 5 to 8 years in central and northern Alaska[33]

Chinook are well known for the prodigious lengths they may travel upstream on spawning runs. There are reports of returning adults finding their way up to 3,000 km in the Yukon River.[33] Returning adults that move into freshwater during the fall tend to spawn in mainstem or tributary streams shortly after they reach the spawning grounds. Those which make their runs into freshwater in the winter and spring may remain in deep pools near the spawning grounds for as long as 5 months before their eggs ripen and spawning activity is initiated.[15]

Female spawners typically select their nesting sites in gravel at the lower lip of a pool just above the riffle.[40,41] The redds average about 6 m^2 in size.[40] During the act of spawning, as many as 10 to 12 males may attempt to mate with a single female by the time all of her eggs have been deposited.[41,42] After spawning, the adults begin to deteriorate rapidly. They exhibit large open wounds and heavy fungal infections. Death normally occurs within 2 to 4 weeks.[41]

Fecundity varies from one chinook stock to the next. For example, in the Sacramento River, the average number of eggs per female is in excess of 7,000, while in the Klamath River, the average is just over 3,600.[36] Some individual females may produce up to 20,000 ova.[42] Eggs range in size from 6.3 to 7.9 mm in diameter,[43] and weigh 0.35 to 0.40 g.[44] The eggs, which are particularly vulnerable to shock,[36] will hatch over a range of 4 to 16°C, though colder temperatures can be tolerated during the later stages of development.[17,30,45] Spring chinook usually spawn when temperatures are declining. Their spawning occurs over a range of 4.5 to 18°C,[46] while fall chinook usually spawn over a temperature range of 5.0 to 13.4°C.[33] Hatching requires 204 days at a constant temperature of 1.7°C and only 28 days at 18.1°C.[47] The time required for hatching is equivalent to roughly 900 to 1,000 thermal units[48] (there is one thermal unit for each degree C above freezing for a 24-hour period).

Following hatching, the fry spend several weeks in the gravel before emerging. The alevins become positively phototaxic when they begin to migrate up out of the gravel.[36] Emergence generally occurs at night[36] in the period from February through June and will vary as a function of latitude, temperature, and the time at which the eggs were deposited.[33] An exception to the spring emergence of fry is the Sacramento River where winter chinook spawn during the spring and the fry hatch and emerge during mid-summer.[33]

Overwinter losses of juvenile chinook salmon (as well as juvenile coho salmon and steelhead trout) have been attributed to stranding and freezing, low dissolved oxygen, and

FIGURE 6. Rainbow trout (*Oncorhynchus mykiss*) immature form. (Computer-generated drawing by Kenneth Adkins, School of Fisheries, University of Washington, Seattle.)

predation.[49] Juveniles can survive short term exposures to 3 mg/l of dissolved oxygen at ≤5°C, but their optimum levels are >9 mg/l at <10°C and 13 mg/l at ≥10°C.[33]

Juvenile chinook salmon have been divided into two groups depending upon their time of migration to seawater. One group migrates early in its first year of life, while the other does not migrate until they have overwintered in freshwater.[50] Fall chinook typically migrate during their first year,[51,52] while spring chinook migrate as yearlings.[53] Winter chinook, which emerge during the summer, migrate during the fall at an age of 4 to 7 months.[54]

Most chinook spawn at 3 to 4 years of age, though older fish are sometimes found. Two-year-old precocious males, known as jacks, are common. There are even reports of mature yearling males which have never migrated to the ocean.[55,56]

F. RAINBOW TROUT (*ONCORHYNCHUS MYKISS*)

The rainbow trout (formerly *Salmo gairdneri*) is one of the most popular coldwater sportfishes in North America. Native to the streams of the Pacific coast,[57] it has been introduced around the world into regions where climate and water quality conditions will support its growth and survival. Sea-run rainbow trout are known as steelhead. Rainbow trout exist in reproducing populations in some lakes and streams in the United States, but significant stocking programs are present in most states where fisheries for the species exist. Characteristics of the species are as follows:[5]

- Fins — dorsal fin with 10 to 12 rays
  caudal fin with shallow fork
  anal fin with 8 to 12 rays
  pectoral fins each with about 15 rays
  pelvic fins each with about 10 rays
- Gill rakers — 17 to 21 on the first arch
- Scales — 115 to 161 above the lateral line

Rainbow trout (Figure 6) are metallic blue on the dorsal surface, and silvery on the side. They have black spots on the dorsal and caudal fins, and along the back. Spawning males have a pink or red band on their sides.[5] Steelhead are steel blue on the back and silver on the sides and belly but take on the colors of nonanadromous rainbows when they enter freshwater.[58] While now classified in the same genus as the Pacific salmon, rainbow trout differ from other members of the genus *Oncorhynchus* in one significant way. Rainbow trout often survive after spawning and may spawn in subsequent years. Many steelhead die following spawning, but there is a significant percentage that survives. The incidence of repeat spawning steelhead decreases from south to north along the Pacific Coast[59] and also varies from stream to stream.[59,60] Repeat spawners do not show any significant size increases between years,[60] and there are few fish which return to spawn more than three times.[61] When they enter freshwater, they rarely eat.[62]

The geographic range of the steelhead is from northern Baja, California to the Bering Sea and Japan. Adults in the Pacific southwest generally weigh less than 4.5 kg, though individuals weighing as much as 11 kg have been captured.[58] Larger steelhead are common in the Pacific Northwest.[59] Maximum life expectancy seems to be 8 to 9 years.[61,63]

Most steelhead can be classified as belonging to one of two races.[59,64-67] Winter steelhead enter streams from November through April, while summer steelhead appear in streams between May and the end of October. Portions of both groups may enter freshwater in the spring or fall instead and are named according to the season when they return. In large rivers, steelhead may enter during most of the year. Winter steelhead usually enter freshwater as maturing fish and spawn soon after they appear, while summer steelhead enter freshwater as immature fish and do not spawn for several months.[58] The summer- and winter-run fish do not seem to interbreed, being isolated temporally and spatially,[64,65] though the potential for interbreeding exists, as does the possibility of interbreeding hatchery and wild fish.[68]

Summer steelhead spawn in January and February, while winter steelhead spawn in April and May. Summer steelhead spawn in smaller streams or further upstream than winter steelhead.[65,66] California stocks seem to consist of only winter-run fish, while Washington, Oregon, and southern British Columbia stocks are a mixture of summer- and winter-run fish. Idaho stocks are primarily summer-run fish that migrate up the Columbia and Snake Rivers, while summer-run fish predominate in northern British Columbia and Alaska.[69]

The sex ratio of steelhead that return to the spawning streams tends to be about 1:1 male:female.[19,70] Fecundity varies with fish size and the geographic origin of the fish,[71] with an average of about 2,000 eggs being produced per kilogram of body weight.[72]

Steelhead spawn in cool, clear water that is well oxygenated and has suitable gravel and current velocities.[17] Spawning depths typically range from 0.1 to 1.5 m in currents with velocities of 23 to 155 cm/s and over bottoms with gravel ranging from 0.64 to 12.7 cm in diameter.[73-76]

Steelhead produce redds that may occupy some 5.5 m^2 of stream bottom and have depths ranging from 7 to 30 cm.[77] Spawning typically occurs when the water temperature is between 3.9 and 9.4°C.[78] Spawning may actually cease when there is a sudden drop in stream temperature.[17] The time required for incubation of the eggs varies as a function of temperature. For example, studies have indicated that eggs require 103.5 days for hatching at 3.5°C,[79], 80 days at 5°C[58], and 19 days at 15°C.[58] In the Pacific Northwest, normal conditions lead to an incubation time of 4 to 7 weeks (28 to 49 days).[77] Incubation can occur over a temperature range of about 0 to 24°C, with optimum being about 10°C.[75]

Low dissolved oxygen levels may delay hatching and increase the incidence of anomalies found in the larvae. The level of dissolved oxygen for incubation of steelhead and other anadromous fishes should be no lower than 5 mg/l.[17] When 20 to 25% of the sediment is less than 6.4 mm in diameter, fry survival and emergence are impaired in both steelhead trout and coho salmon.[80,81] Gravel mixtures with high percentages of fine sediments are undesirable for good steelhead production.[82]

Alevins absorb their yolk quickly and become free swimming in only 3 to 7 days.[77] The fry are often found in small schools which usually form in shallow water along the banks of streams. The schools disperse as the fish grow.[58] Food of the juveniles consists of organisms associated with the stream bottom,[79] such as isopods, amphipods, and insects.[19]

Newly emerged steelhead fry are sometimes consumed by juveniles of the same species.[19,83] Studies in Idaho have demonstrated that steelhead fry are at least as viable as resident rainbow trout fry. Steelhead, in fact, tended to displace resident rainbow trout, but did not appear to affect resident brook trout (*Salvelinus fontinalis*).[84] In general, interactions between rainbow trout and juvenile steelhead do not seem to be great because the two prefer different habitats (rainbow select higher water velocities and deeper water than steelheads).[85]

Wild juvenile steelhead commonly spend 2 or 3 years in freshwater,[77] though they may stay as long as 4 years in freshwater.[58] Hatchery fish migrate to sea after only 1 year in freshwater.[77] During their freshwater phase, steelhead may be preyed upon by large rainbow trout, sea-run cutthroat trout (*Salmo clarki*), sculpins (*Cottus* spp.), great blue herons (*Ardea herodias*), mergansers (*Mergus merganser*), and various mammals.[69]

The fish remain in the ocean from 1 to 4 years before returning to spawn,[58] with the length of time spent in both freshwater and the ocean generally increasing from south to north along the Pacific Coast of North America.[59] Age groups are often designated in a manner which indicates the amount of time which the fish spent in freshwater and in the ocean. Thus, a fish designated 2/3 would have spent 2 years in freshwater and 3 summers or winters in the ocean.[62]

Smolting and seaward migration in steelhead, as in other anadromous salmonids, are initiated by various environmental factors including photoperiod, water temperature, and water chemistry.[27,86] The minimum size at which smolts can successfully enter seawater seems to be from 14 to 16 cm.[87,88] Smolts planted in a particular stream will usually return to that stream as adults,[89] and while the homing instinct in steelhead has never been precisely explained, a learning process called imprinting appears to be involved.[90] Low temperatures have been associated with extended periods of smolting, while the process is accelerated when temperatures are warmer.[91, 92]

While living in the marine environment, steelhead consume squid, amphipods, and juvenile greenling (*Hexagrammos* spp.)[93,94] The primary predators on steelhead at sea are other predatory fishes and marine mammals.[77]

G. CUTTHROAT TROUT (*SALMO CLARKI*)

The cutthroat trout, like the rainbow, has both land-locked and anadromous strains. Both the nonanadromous and the coastal or sea-run cutthroat trout are of interest to aquaculturists, particularly in terms of production for enhancement stocking. There is no commercial production of cutthroat as a foodfish in the United States at the present time.

Coastal cutthroat trout (*O. clarki clarki*) are distributed from northern California to southeastern Alaska and are rarely found more than 160 km inland.[79,95] An intermountain, or western slope, subspecies (*O. clarki lewisi*) has also been recognized,[79] which has been introduced to the Cascade Mountains and the waters of central and eastern Washington. Characteristics of the cutthroat trout (Figure 7) are as follows:[5]

- Fins — dorsal fin with 8 to 11 rays
  caudal fin with shallow fork
  anal fin with 8 to 12 rays
  pectoral fins each with about 13 rays
  pelvic fins each with about 9 rays
- Gill rakers — 15 to 22 on the first arch
- Scales — 120 to 180 above the lateral line

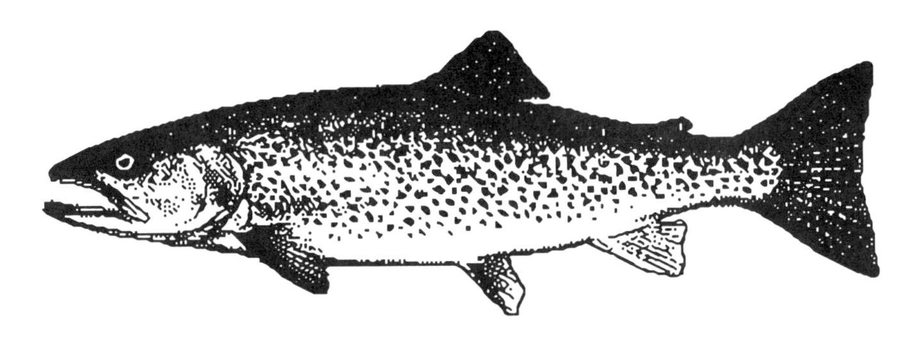

FIGURE 7. Cutthroat trout (*Salmo clarki*) adult. (Computer-generated drawing by Kenneth Adkins, School of Fisheries, University of Washington, Seattle.)

Cutthroat trout are greenish blue on the dorsal surface and silvery on the sides.[5] They have angular or round black spots on the sides and on the vertical and pectoral fins. There is a red or orange streak along the inner edge of the lower jaw in fresh specimens.[5,95]

The inland subspecies have been known to reach lengths of about 1 m and may weigh as much as 18.6 kg,[79] though 8 kg or less is more common. The coastal subspecies, by contrast, only reaches about 1.8 kg.[5]

Inland cutthroat spawn in April and early May, with the peak occurring near mid-April. Sea-run cutthroat typically spawn from late December to February,[79] with some geographic variation.[95]

The time during which sea-run cutthroat return to freshwater for spawning is fairly constant within a given stream but varies from one stream to another.[96,97] For example, in Oregon and Washington, freshwater entry occurs from July through March,[98,99] and few of the fish overwinter in seawater.[97] The run in Alaska is limited to the spring.[100] In Alaska, only a small number of the returning fish are immature,[97,101-105] while in the Columbia River, Puget Sound, and British Columbia a large percentage of the females returning for the first time to freshwater do not spawn.[97]

Young cutthroat trout typically remain in small headwater streams for a year before migrating to large rivers where they may remain for variable periods.[96,106-108] In Washington and Oregon, the majority migrate to sea at 3 years of age, while the average in Alaska is about 6 years.[79] Cutthroat school before entering the marine environment and remain schooled during the saltwater migratory period.[109] Migrations to sea occur from January through June, with the peak in April through June. Prior to smolting, parr may migrate up- and downstream several times.[106,110] The fish return to coastal estuaries beginning in July and extending through January[79] of the same year during which they migrate to sea.[96,99,104,109] Returning fish move toward the headwaters of their natal home stream tributaries[95] and occur sympatrically with resident cutthroat populations.[97,111]

Spawning redds are typically about 60 cm long and 45 cm wide.[79] The fish generally spawn in the tail ends of pools at depths as shallow as 10 to 15 cm[112] or as deep as 1 m.[96] Some fish spawn in riffle areas.[79,112] The eggs are laid in a wide range of gravel sizes at depths of 13 to 18 cm.[79] Spawning gravel size ranges from 0.16 to 10.2 cm.[74,113] The eggs, which range in number from a few hundred to a few thousand, depending upon the size of the female,[79,95] are from 4.3 to 5.1 mm in diameter.[95]

Cutthroat trout are repeat spawners, but mortality related with spawning is high.[20,98,109,114] Maximum lifespan is about 10 years,[104] with most adults in a typical run ranging from 3 to 6 years of age.[96,115] Surviving adults enter seawater earlier in the year than smolting juveniles.[103,109] Alevins remain in the gravel for 1 to 2 weeks following hatching and then emerge as fry which are less than 3 cm long. Fry typically inhabit shallow, low-velocity areas at the stream margins.[96]

Spawning occurs over a temperature range of 6 to 17°C,[74] with optimum in the range of 9 to 12°C.[78] Optimum temperature for egg incubation is 10 to 11°C.[116-117] The fish are usually not found where maximum temperature is consistently above 22°C,[95] though they can tolerate brief periods of exposure to 26°C if such exposures are followed by considerable subsequent cooling.[118] Culturists should be aware that dissolved oxygen levels below 5 mg/l should be avoided during the summer months.[119,120]

In freshwater, cutthroat trout feed primarily on aquatic insects, though the larger individuals are primarily piscivorous.[79] Sea-run strain cutthroat seem to be more opportunistic in freshwater than their landlocked cousins. The former consume not only insects, but also zooplankton, frogs, earthworms, juvenile salmonids, crayfish, fish eggs, terrestrial insects, and salamanders.[20,74,79,119,121-132] During the time cutthroat trout are in the marine environment they feed on gammarid amphipods, isopods, callianassid shrimp, immature crabs, and various types of fishes including chum salmon, pink salmon, Pacific sand lance (*Ammodytes hexapterus*), and others.[133,134]

Predators on cutthroat fry in freshwater include rainbow trout, Dolly Varden (*Salvelinus malma*), shorthead sculpins (*Cottus confusus*), and adult cutthroat. Other predators are blue herons, belted kingfishers (*Ceryle alcyon*), and mink (*Mustela vison*).[135] In the marine environment cutthroat are subject to predation by Pacific hake (*Merluccius productus*), spiny dogfish (*Squalus acanthias*), harbor seals (*Phoca vitalina*), and adult salmon.[109]

H. ATLANTIC SALMON (*SALMO SALAR*)

The Atlantic salmon is found on both sides of the Atlantic Ocean. It occurs from the Connecticut River northward to Ungava Bay on the east coast of North America,[5] and can be found in Iceland and Greenland.[5,136] The species occurs in the British Isles and from northern Spain to northern Norway in Europe.[5,136] Populations have been heavily impacted by fishing and industrialization in the northeastern United States, but efforts are being made to reestablish runs of the fish. Nonmigratory populations of Atlantic salmon have been found in the U.S.S.R., Norway, Sweden, and Finland, but most populations are anadromous.[136]

The fish is light brown on the dorsal surface, silvery on the sides, and has black spots on the dorsal, adipose, and anal fins (Figure 8) that may be obscured in animals taken from saltwater. The following characteristics distinguish the species:[5]

- Fins — dorsal fin with 11 to 12 rays
  caudal fin moderately forked
  anal fin with 8 to 12 rays
  pectoral fins each with about 14 rays
  pelvic fins each with about 10 rays
- Scales — about 120 along the lateral line

Maximum size of the Atlantic salmon is 27 kg and 122 cm.[5] The fish survive a temperature range of about 1 to 24°C, with an optimum of 10 to 17°C.[45]

As is true of other anadromous salmon, Atlantics return to their natal stream for spawning. Like steelhead trout and sea-run cutthroat trout, Atlantic salmon may survive after spawning, though a relatively high rate of mortality subsequently occurs. The typical spawning season is from October through January, though some stocks spawn in February and March.[136]

Females construct redds into which they lay from about 1,200 to 2,000 eggs per kg of body weight. Spawning occurs in waters of 6 to 10°C,[45] and incubation requires about 440

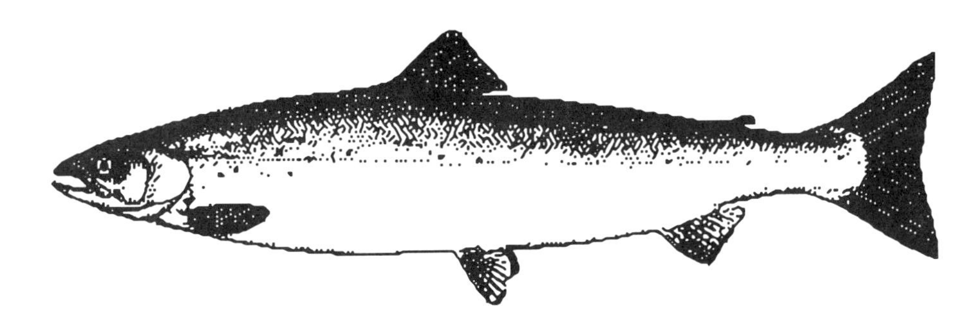

FIGURE 8. Atlantic salmon (*Salmo salar*) immature form. (Computer-generated drawing by Kenneth Adkins, School of Fisheries, University of Washington, Seattle.)

thermal units.[136] Yolk sacs are absorbed by the alevins over a period of 3 to 4 weeks, depending upon water temperature, after which the fish emerge and begin feeding. Fry feed on zooplankton. Parr are able to take larger food organisms.

Parr Atlantic salmon are often found in shallow riffles. At the southern end of their range, the parr-smolt transformation may occur when the fish are slightly over 1 year of age, while in the northern extremes of the range, fish may be 5 or 6 years of age when they smolt.[136]

After migrating to sea, the smolts feed on amphipods. As they grow, their food habits change toward piscivory, though krill (Euphausiidae) and Arctic prawn (*Pandalus borealis*) may also be consumed. Common fishes consumed by Atlantic salmon at sea include capelin (*Mallotus villosus*), sandeels (*Ammodytes* spp.), and members of the herring family (Clupeidae).[136] The time that Atlantic salmon remain at sea may range from a few months to 2 years.[137]

As is true of various other salmon species, the production of precocious males, or jacks, often occurs with respect to Atlantic salmon. Early maturation appears to be related to both genetic and environmental factors, and under favorable growth conditions it often occurs in the most rapidly growing males.[138,139] Early transfer of juvenile Atlantic salmon to seawater has been shown to reduce the incidence of jacking.[140]

Atlantic salmon have been introduced to the Pacific coast of the United States, but no known reproducing populations have been established.[141] With the recent development of net-pen Atlantic salmon culture on the west coast of North America, concern has been raised about escapement and interbreeding of Atlantic and Pacific salmon species. Attempts to produce hybrids between Atlantic salmon and various species of Pacific salmon have failed to yield viable embryos, and it has been concluded that crosses between those species could not occur in nature.[142,143]

I. BROWN TROUT (*SALMO TRUTTA*)

Brown trout (Figure 9) were introduced into North America from Eurasia in 1883 and are widely distributed throughout the trout fishing regions of the United States.[144] The species is a highly valued sportfish in many regions. The average size is usually between 0.1 and 1.8-kg, though the record is an 18-kg specimen taken in Scotland.[5,144] Recognition characteristics of the brown trout are as follows:[5]

- Fins — dorsal fin with 10 to 11 rays
  caudal fin with shallow fork
  anal fin with 9 to 12 rays

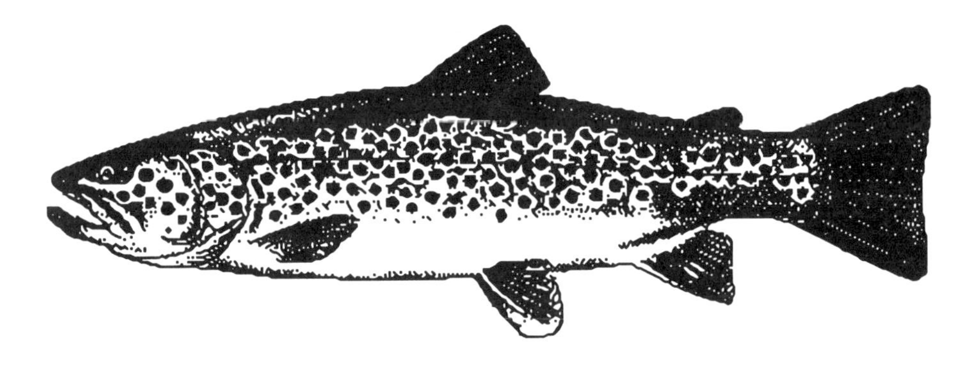

FIGURE 9. Brown trout (*Salmo trutta*) adult. (Computer-generated drawing by Kenneth Adkins, School of Fisheries, University of Washington, Seattle.)

pectoral fins each with about 13 rays
pelvic fins each with about 9 rays

- Gill rakers number 16 to 19 on the first arch
- Scales — from 116 to 130 above the lateral line

Brown trout are nonanadromous fishes which spawn in the fall. The fish can survive a temperature range of about 1 to 26°C, with optimum occurring in the range of 9 to 16°C. Brown trout spawn over a temperature range of 9 to 13°C.[45] They feed primarily on terrestrial and aquatic insects, though crustaceans and fish become increasingly important food items as the fish exceed 25 cm in length.[144]

IV. HOMING IN SALMONIDS

One of the most intriguing and vexing problems associated with anadromous salmonid biology has surrounded the long-recognized phenomenon that the fish return, almost without exception, to spawn in the same stream in which their parents spawned. There is, to be sure, some straying from the home stream by individual fish; if that were not true, races, subspecies, and even entire species of salmon would have been lost. Salmon, particularly Pacific salmon, tend to live in areas which feature high levels of tectonic activity. That activity has often led to major changes in land configuration which may involve relocation of stream channels, creation of new streams, and elimination of old watercourses. Straying by salmon allows populations to survive displacement from home streams which are no longer accessible and to become established in new river systems.

While straying does occur, the vast majority of the fish returning from the ocean are able to not only find the proper continent, but also find the proper river and its tributaries. Spawning may occur within a few meters of where the individual fishes were incubated when they were at the egg stage.

Studies led by Arthur D. Hasler, a University of Wisconsin biologist who conducted some of his research in the Pacific Northwest, led to the olfactory hypothesis to explain homing in salmon. The hypothesis, which attempts to explain the latter portion of the homeward migration of salmon (that is, the portion which relates to finding the home stream), attributes the phenomenon to olfactory cues that are specific to each location in the stream and are learned by the juvenile fish shortly before they migrate to sea.[145-148] There are three basic tenets to the hypothesis:[146]

- Because of local differences in soil and vegetation within a given drainage basin, each stream will have a unique chemical composition, and thus, a distinctive odor
- Before juvenile salmon migrate to the sea they become imprinted in terms of the distinctive odor of their home stream
- Adult salmon use the imprinted information as a cue for homing when they migrate through the home stream network and make their way back to the home tributary

Thus, memory of the home stream is not inherited, but is associated with a period of irreversible learning during which the fish learns to recognize the home stream from its distinctive odor.

Lauren R. Donaldson used the hypothesis to good advantage when he created an artificial salmon run on the campus of the University of Washington. Donaldson constructed a pond which receives pumped water from an adjacent lake. Salmon of various species are held in the pond for a few weeks before release and, after variable periods at sea, will subsequently return to the pond to spawn. On their return the fish pass around a set of locks by climbing a fish ladder; then they must swim several kilometers through a heavily industrialized lake that is virtually surrounded by the city of Seattle. The fish are then able to distinguish that the water which exits from the spawning pond through a short fish ladder is distinctive enough to signify that they have reached their ''home stream''. It is assumed that odors associated with the gravel in the pond are the key since the water is pumped into the pond from the adjacent lake.

An alternate explanation has been proposed,[149,150] in which the primary cue to home stream recognition is hypothesized as being attributable to population-specific pheromones which guide homing adult salmon. Examination of the latter hypothesis has led recent investigators to the conclusion that while conspecifics do seem able to recognize one another, that recognition plays no important role in the homing phenomenon. Learned odor recognition and not pheromones seems to be the most reliable explanation.[151-153]

Migration of salmon in offshore waters appears to be distinct from that in freshwater and has not been nearly so well explained. Hypotheses for orientation of salmon in the ocean have included the involvement of magnetic and celestial information.[154,155] Thomas P. Quinn has proposed a high seas salmon migration model which includes a map-compass-calendar system.[155] The proposed map relates magnetic inclination and declination on a bi-coordinate grid. Compass mechanisms could include both celestial and magnetic orientation. The fish would also need to have the ability to assess changing daylength in combination with an endogenous circannual rhythm sense to provide an internal calendar. While no hypothesis for high seas navigation of salmonids has received widespread support to date, it is felt that the strong orientation of salmon in the open ocean must be explained by any hypothesis that is proposed.[156]

The last word on salmon migration has certainly not been heard. The subject continues to be of significant interest and has been the subject of at least two symposia.[157,158]

V. PARR-SMOLT TRANSFORMATION IN ANADROMOUS SALMONIDS

The physiological changes, known as the parr-smolt transformation or smoltification, which prepare salmon for the transition from freshwater to seawater, have been the subject of a great deal of study by salmonid researchers.[86,159] Physiologically, the changes associated with smoltification in chinook salmon appear to be more complex than in coho salmon and rainbow trout.[160,161] Smolted fish will tolerate high levels of chloride and readily adapt to high salinity.

From a morphological standpoint, smoltification is associated with the disappearance of the dark pigmented parr marks which are characteristically found on the sides of young salmon. As the fish become more silvery, they also become increasingly tolerant of salt water and will normally begin their migration to the marine environment at that time.[159] Recognition of smoltification is extremely important to net-pen culturists who artificially transfer their fish from freshwater to saltwater net-pens, typically without acclimation, when the fish are judged to be smolts. Errors can lead to osmoregulatory stress and high fish mortality.[162-166] Mortalities of from 25 to 50% or reversion to saltwater parr may occur when fish are transferred to net-pens prematurely.[164,167,168]

Smoltification has been shown to be associated with fish size in salmon and anadromous trout,[169-174] as has age.[175] The minimum size for successful transfer of coho salmon from freshwater to saltwater is 15 g,[176] though a minimum size at transfer of 20 g[177] or even 40 g[168] has been recommended. A strong genetic relationship between size and various traits which characterize smoltification, with the exception of percent survival, has been observed in coho salmon.[174] Environmental factors related to the onset of the parr-smolt transformation include photoperiod and temperature.[178-180] Water of warmer than normal temperature has been used to promote smoltification in coho salmon,[25] as has modifying the amount of light available to Atlantic salmon.[181]

Thyroid hormones appear to control the metabolic and developmental changes that occur during smoltification, and changes in the levels of circulating thyroid hormones have been observed during the parr-smolt transformation.[182-185] The thyroid hormone T_3 has been shown to produce positive growth responses in coho and chinook salmon as well as rainbow trout[186] and one possible way to promote smolting would be to add that thyroid hormone to the diet of presmolts.[187] Survival of zero-age coho salmon in seawater seems to be correlated with the progressive elevation of plasma T_3, while yearling success in seawater is related to the T_4 hormone peak.[188,189] The thyroid endocrine system is depressed when salmonids are transferred directly from freshwater to seawater.[190]

Increasing levels of ATPase activity have been found in association with smoltification in salmonids by a number of investigators. In coho and chinook salmon, as well as steelhead trout, gill Na^+,K^+-ATPase activity increases during smolting,[191-194] and increased plasma levels of Na^+,K^+-ATPase have been found in coho salmon.[189] In addition, during smoltification there is a proliferation of chloride cells in the gills and subopercular epithelium, as well as involvement by various endocrine systems.[86]

February gill ATPase activity was found to be more elevated in Atlantic salmon reared at ambient temperatures than those which were exposed to elevated temperatures,[195] and smolt survival was higher in the ambient temperature fish. This led to the conclusion that elevated winter temperatures may be detrimental to subsequent seawater survival of smolts.

Chapter 2

RAINBOW TROUT CULTURE

Ernest L. Brannon

TABLE OF CONTENTS

I. INTRODUCTION

Rainbow trout (*Oncorhynchus mykiss,* formerly *Salmo gairdneri*) can be classified as the premier salmonid species of the world. They are found in saltwater as large and silvery as their Pacific cousins, or resident in freshwater with scarlet sides and as spotted as *Salmo* or *Salvalinus.* As a sportfish rainbow trout are an angler's delight, breaking the surface from dark holes in mountain streams for a skipping fly, or thrashing through whitewater to challenge the strongest line. Although endemic to the western region of North America, from the Rocky Mountains west to the Pacific Coast and from the Rio Casa Grande north to Norton Sound,[196] rainbow trout are the most widely cultured species in the world. Because they tolerate water temperatures from 0 to 27°C, their eggs have a short incubation period, the fish grow fast and eat a variety of items as surface feeders, and the species has been transferred around the world as a highly popular fish for both sport and commercial use. They are now found on every continent except Antartica, and occur as far north as Finland, in the mountains of equatorial regions, on major islands like New Guinea, and south to New Zealand.

Rainbow trout are unique and are the most forgiving salmonid to displacement and abuse from our management practices. That tolerance has been responsible for the rather homotaxic form that characterizes rainbow trout maintained by our conservation hatcheries as brood stock. The criteria we have used in developing hatchery stocks for conservation purposes have been similar to those for commercial hatcheries. Most conservation hatchery brood stocks are better suited for commercial purposes than for supplementing natural populations, which gets to the dilemma in which we find ourselves. Selection programs to develop fast-growing or highly fecund stocks adapted to intensive pond culture will not be the appropriate fish for enhancement purposes. Stocks used for put-and-take fisheries may very well be commercially raised trout, but enhancement objectives will be very different. This difference in how hatchery fish should be employed has often been overlooked and is the major reason why hatcheries are viewed disparagingly by some conservationists. The fish we use and the technology we employ have to be determined by the objectives sought. This chapter examines the life history characteristics of rainbow trout, and presents some of the hatchery technology that may be helpful in developing culture strategies.

II. NATURAL LIFE HISTORY

Rainbow trout were recently reclassified as a species in the genus *Oncorhynchus* with the Pacific salmon. While it may now be legitimate to call them ''salmon trout'', they are still the Pacific cousin of *Salmo salar.* In their anadromous form they resemble Atlantic salmon more than others in the genus *Oncorhynchus.* Body proportions, tail shape, silvery sides, and spot pattern appear much as they do in Atlantic salmon. Taxonomic traits, such as fin ray counts, place the rainbow trout with the genus *Salmo.* Rainbow trout can spawn in multiple years like *S. salar* and, unlike the hooknose characteristic of *Oncorhnychus,* at maturation the lower jaw of the rainbow hooks the same as that of *S. salar.*

Anadromy evolved early in the phylogeny of the family before the genus *Oncorhynchus* separated from *Salmo* and was the feature that allowed their common ancestor to multiply its diversity and colonize so many different habitats.[197] Anadromy provided the selective advantage to access food resources for greater reproductive success. They could retreat to the freshwater environment for comparatively secure spawning and incubation conditions, and then escape to sea to avoid the limited growth often imposed by the low productivity of that environment. Or, if conditions were suitable, they could remain in freshwater for their entire life. Rainbow trout are most highly successful with this dichotomy in life history. They will assume whichever life history strategy that affords them the greatest advantage.

When access to the sea is possible, rainbow trout can be found as anadromous steelhead, and make 3-, 4-, and 5-year marine treks from the Pacific Northwest as far as the Gulf of Alaska to indulge in rich stores of plankton and small fish produced there. Steelhead return at sizes ranging from 2 to 18 kg, historically migrating as far as 1,700 km up the Columbia and Fraser Rivers to spawn. In contrast, where access to sea is prevented or where freshwater resources are sufficient, they can equally well assume a freshwater resident form and live their life successfully in numerous niches found over their range. In high mountain lakes or streams, the fish will manifest characteristics that suit the sparse productivity of their habitat and may not exceed the size of 15 cm and 40 g at maturation, with a fecundity of less than 100 eggs. In large productive lake basins, however, rainbow trout may approach sizes of 100 cm and 16 kg, with a fecundity of several thousand eggs.

The characteristic that sets rainbow trout apart from salmon and gives the species a more flexible strategy is its spawning time. Most salmonids are fall spawners, and their progeny incubate during the fall and winter months to emerge in synchrony with the spring bloom in the food supply. Based on the temperature of the incubation stream and the length of their temperature-dependent incubation period, the appropriate emergence time dictates quite precisely when spawning must occur to assure that synchrony. Spawning among the salmonids, therefore, is timed to occur proportionally earlier in cool incubation streams and later in warmer streams. But in contrast to the fall spawning time of their relatives, rainbow trout are considered spring spawners. This temporal window allows them to avoid competition for space with their often numerous relatives, and more importantly, it allows them to use habitat that is inhospitable or nonexistent during the winter months, such as in high elevation streams. Spring time spawning gave rainbow the ability to colonize habitat not available to salmon.

By evolving such a spring spawning strategy, however, rainbow trout have had to greatly accelerate their incubation period to still synchronize emergence with the spring food supply. Rainbow trout incubation occurs in half the time required by their fall spawning relatives at any given temperature, but the exact time of spawning is still site specific. Repetition of such site-specific behavior is assured by incorporating genotypic timing attributes.[198] Summer-run steelhead will enter river months before spawning, using the higher flows to reach the cool spawning stream they seek, after which they will wait in holding areas through low winter stream discharge until the proper spawning time approaches in the spring. Winter-run steelhead, with access unrestricted by flow in their warmer natal streams, will enter freshwater just weeks before spawning. The same variability is found in all of the resident rainbow stocks. In high elevation lakes, some stocks spawn late in June because access to the streams is not possible until the ice leaves, while in warmer lowland lakes, stocks are predisposed to spawn in March or April. In each case, stocks are site specific in their timing characteristics to maximize survival.

Patterns of orientation during migration and distribution throughout the entire life cycle of rainbow trout and other salmonids have a strong genetic basis.[154,199,200] Behavior as basic as the time of readiness to leave a particular stream, and as broad as following an ancestral migratory path over several months at sea, is strongly influenced by the genetic predisposition of the stock. Homing behavior involves the ability of adults to retrace their pathway experienced as juveniles when leaving the incubation site and nursery stream. Homing evolved to assure that the genotype returns to a certain habitat, and it has to be precise or all of the cost invested in genetic specificity would be rendered useless.

Application of life history is critical to the development of management strategies with hatchery fish. Management programs that disregard the genetic background of the stock, or the sequence of events necessary for good homing success, will result in fish being out of synchrony with the environment, and result in long-term fitness losses with implications more severe than overfishing. It is important that managers preserve the biodiversity inherent in fish populations. Even when the environment has changed from that in which a stock evolved, or when a stock no longer represents an appropriate phenotype, future management options will need access to genetic diversity to create new populations that selection can rework. Fisheries management using hatchery fish must operate within the guidelines that regulate natural populations, or the public will have to be content with sustaining the hatchery programs that have replaced them.

III. ARTIFICIAL PROPAGATION

Culture of rainbow trout, as with other salmonids, is over 100 years old in the United States. The old *U.S. Fish Manual*[201] on the artificial propagation of salmonids of the Pacific coast was printed in 1903 and, with some exceptions, doesn't differ markedly from the techniques used today. The greatest advances in present day trout culture have been primarily in nutrition and pathology. Although most of the biological concerns raised against public hatchery fish are justified, if it were not for hatchery technology, rainbow trout would be in a desperate situation over much of their range. Hatcheries have sustained rainbow trout in the Pacific coast states and to some degree in Canada. In the northern range of the species in British Columbia and Alaska, the habitat of the rainbow remains unaltered to a large extent, but forest management practices have encroached severely on many streams there as well as in more southern locations.

Artificial propagation started with egg incubation facilities and unfed fry plantings. Early biologists noted that mortality associated with natural incubation in streams resulted in survivals to the fry stage of less than 5% of the eggs spawned. They concluded that artificial measures could greatly improve incubation success. This was true, and the early incubation stations succeeded in achieving survival rates over 90%. Adult returns from those plants, however, were not proportionally better. Life history requirements of the fish were not well understood, and densities planted had no relationship with the carrying capacity of the streams. Poor fry survival, and later fingerling survival, were used as justification to rear fry to the smolt stages to eliminate sources of high mortality, but return survival was often less than 0.01%.

A major problem with hatchery propagation was the feed used for the fish. Slaughterhouse waste and salmon carcasses were a major source of feed for cultured salmonids. Nothing was known about vertical transmission of disease from uncooked carcasses to young fish, and as a result, tuberculosis, bacterial kidney disease, and other serious problems resulted from such practices. With the pioneering work on fish nutrition by investigators with the United States Fish and Wildlife Service, and with pasteurization of feed ingredients, hatchery successes made great strides forward. Rainbow trout were the first species for which nutritional requirements were determined, and diet formulations resulted that improved culture success. Conservation hatcheries have had return rates over 10% of the steelhead released, but much more work is still required.

Rearing facilities suitable to hold fish until release also changed. The original earthen ponds that were common in early hatcheries gave way to reinforced concrete circular ponds, but those still had inadequate outlets and flow patterns to assure sufficient flushing. As state and federal hatchery programs expanded, and especially with multimillion dollar mitigation programs, large sums of money were invested in pond and water intake systems. Conservation hatchery costs rose to staggering levels, exceeding tens of millions of dollars, amounts made possible only because they were justified as mitigation for the loss of fish habitat by irrigation and hydropower development.

Long before the 1990s, conservation hatchery costs moved away from the type of facilities commercial hatcheries could afford or even justify. Dirt ponds are still in use on commercial farms, and where concrete raceways are employed, their costs are an order of magnitude below the expenditures in state or federal hatchery programs. The outstanding examples of advanced commercial facilities are the private trout hatcheries in Idaho. Some have spring water flows as high as 8,500 l/s and produce 6 kg for each liter per minute of water flow. Some are vertically integrated companies, operating their own hatcheries, feed mills, processing plants, transportation networks, marketing programs, and research facilities.

Commercial hatcheries are leading in much of the technology used in fish culture. Feed distribution mechanisms, oxygen recharge, pond cleaning operations, waste treatment, and water reuse are as advanced in private facilities as they are in conservation hatcheries. Larger commercial hatcheries extract their pond wastes, settle solids, and recondition the water before reusing it in the next raceway. Some of the fish farms pump pond effluent waste directly onto crop land as fertilizer through sprinklers, and others distribute the nutrient-rich solids.

In some respects, hatchery technology has made a complete circle. Unfed fry releases are used in some of the most recent enhancement programs being developed for rainbow trout. While fry releases are much more common among salmon hatcheries, especially the "non-profit" pink and chum salmon hatcheries in Alaska, trout fry release projects are included on many small streams by sportsmen and conservation groups such as "Adopt-a-Stream" in Washington State.

Rainbow trout fall into that category of salmonids, with coho and chinook salmon, referred to as stream dwellers. Stream dwellers are limited by the food productivity of their stream, and care must be taken to not exceed the natural rearing capacity. Spawners in stream-dwelling populations usually provide more than enough gametes for the carrying capacity of the stream. In contrast to species such as pink, chum, and sockeye salmon that only use the stream for incubation before migrating to lake or marine feeding areas, state-of-the-art technology such as spawning channels, mass artificial incubation beds, and large hatchery fry release programs are not appropriate technology for rainbow trout. Enhancement programs for rainbows involve the release of relatively small numbers of fry that can be easily accommodated in gravel incubation boxes installed streamside and where fry volitionally release themselves at emergence. Gravel box incubation represents the most simple type of hatchery, but is an example of the best technology to apply in some circumstances.

Appropriate technology is the key to success of any artificial propagation program undertaken. Operations need to integrate technology based on the requirements of the fish being released. The following section presents some of the concepts and practices associated with hatchery operations and compares conservation and commercial hatchery operations to emphasize different procedures that should be followed at such facilities.

IV. HATCHERY TECHNOLOGY

There is a natural division in the culture strategy for rainbow trout between conservation and commercial hatcheries. Hatchery propagation is essential to both, but their production objectives are very different, and therefore the manner in which culture techniques are employed has to be different. Conservation hatcheries are enhancement tools. In some situations, such as where major dams deny access to habitat, complete dependence on hatcheries to maintain the runs may be required. In most other situations hatcheries should play only a supplementary role, increasing the production of native fish by providing higher survival success to the run, and preferably never becoming a major production center, nor altering the genetic integrity of the population.

Commercial hatcheries are production centers, maximizing the efficiency of each unit of water, with fish density and source of stock dictated simply by the market targeted. Genetics have a different role, tailoring the commercial product to increase growth rate, maximize dressed weight, and improve disease resistance. These attributes have good commercial value, but most likely would not be desirable traits to breed into native populations. Sportsmen often do not appreciate these differences, and they frequently request state agencies to plant king-size strains from hatchery populations into their local favorite fly-fishing streams. We must not overlook the fact that if such traits had superior survival value under natural environmental circumstances, they would likely have evolved. More important, however, if they do not occur naturally they will not be sustained when introduced in a new environment. On the other hand, commercially desirable traits can be sustained in cultured populations because the culture environment artificially maintains the selective pressures through breeding programs. Under hatchery conditions selective pressures can be controlled and the desired requirements can be satisfied.

A. CONSERVATION HATCHERY BROOD STOCK

Fitness, defined by Falconer[202] as the proportionate contribution of offspring to the next generation, is critical in the identification of brood stock for conservation hatcheries. In any particular stream, the characteristics of the environment and the native fishes have co-evolved. Populations of rainbow trout resident in a stream, or steelhead returning to a river system, are not homogeneous, but rather are made up of several subpopulations that return to unique sites or tributaries within the system. These small populations have differences that translate into spatial or temporal traits inherent to their fitness. When fish are removed from their natural habitat, fitness will decrease in proportion to the degree of separation from their environmental requirements. Salmonid life history traits such as spawning time, emergence behavior, orientation, distribution, and migratory patterns are based on genetic components.[197,203,204] It is critical that enhancement programs by conservation hatcheries take measures to assure that such populations traits are not lost. There is no alternative, therefore, to the use of spawners that originate from the target stream or other location in the watershed, and the progeny of the brood stock should be returned to the same location when released.

Because of high costs involved in the construction of conservation hatcheries, they are often used as production centers, and their fish are distributed to many regional streams. However, if fish are planted without regard for their origin, the synchrony between environment and stock is compromised, and the ability of the populations to sustain themselves is limited. More seriously, the reproductive fitness of the native population is at risk because they will spawn with hatchery fish. Evidence of genetic integration between hatchery and native rainbow trout has been demonstrated in the upper Yakima River, Washington.[205]

Stock degradation occurs even more rapidly by interception of prespawners on their migration upstream in large river systems for hatchery propagation. In large river systems the range of environmental variation can be greater over the length of the river than between tributaries, and the uniqueness of stocks would also be greater. Interception of prespawners, irrespective of where they are going, and redistribution of their progeny would immediately disrupt the continuity between stocks and their environments. Since the young fish imprint on odor cues of the hatchery or site of introduction for homing, they are prevented from returning to match up with their ancestral stream. Examples are the chinook salmon runs in the Columbia River where the early fish intercepted at Bonneville Dam were spring chinook destined for the upper river. Those fish were not the appropriate stock for release in the lower river. Their return time was months too early, and the migratory distance far too short to develop a self-sustaining stock suited for high summer temperatures in the lower river environment.

Of course, when dams block access to the upper reaches of a river system, propagation of upriver stocks in a hatchery may be the only hope of maintaining them. The Cowlitz River hatcheries in the Columbia River system are good examples of where salmon and steelhead populations are sustained by artificial propagation. The Cowlitz hydropower dams blocked steelhead trout and spring chinook salmon from returning to their ancestral spawning grounds at the foot of Mt. Rainier, and two hatcheries were built to mitigate for losses to the upper river. Such a management policy is necessary under such circumstances, but any idea that the procedure will promote natural production of those strains in the lower river and preserve the ancestral traits, is in error. Unless displaced salmonid populations are sustained by the hatchery, if they survive at all, they will be altered by changes conforming to selective forces of the local environment.

Even sustained hatchery production can eventually change the nature of the propagated fish unless care is taken to develop a breeding program to maintain the full range of genetic characteristics of present stocks. Efforts need to be taken to maximize the number of spawners used and to spawn in pairs to maintain as much variability as possible. To avoid using second generation hatchery fish that home and make themselves more accessible, hatchery fish should be marked for recognition. In this way creating a "hatchery population" can be avoided by using only breeding stock from naturally produced fish. The Idaho Department of Fish and Game marks all hatchery steelhead with an adipose fin clip and allows fishermen to keep only marked fish to conserve the wild strains.

Where stocks have been eliminated, and new runs are being developed, it will be rare that any selected replacement will have the attributes appropriate for sustained natural survival. In these cases it would be better not to use an existing stock for replacement, but rather to cross strains that have timing patterns similar to those of the original stock. Crossing different strains will break down the genetic homeostasis maintained within strains and markedly increase the variability on which natural selection can work to establish a new run.

B. COMMERCIAL HATCHERY BROOD STOCK

Rainbow trout brood stock in the commercial hatchery are developed with traits that will improve the economic value of the fish. These values may be unrelated to any ecological significance for the species. Development of commercially desirable or marketable traits is possible through various breeding programs that make use of stock hybridization, genetic selection, or molecular biology. The simplest and most readily applied technique is to hybridize stocks, if by crossing them one will produce the desired commercially important trait. Since the characteristics that generally typify differences between individuals are quantitative or differences involving many loci, the F_1 progeny of such a cross will be intermediate in one degree or another to those traits in the parents. Fish brightness is an example of a trait that will provide an economic dividend. Fast growing, deep-bodied rainbow trout can be hybridized with the silvery bright anadromous form, and salmon bright, well-proportioned progeny will result. Each strain is kept as separate brood stock and the commercially desirable traits are produced in the F_1 generation by crossing strains. Selective breeding for the trait is not necessary. In fact, using the F_1 progeny as brood stock would be undesirable because the whole array of combinations would result in the F_2 generation with only a portion demonstrating brightness. The potential of this approach in improving marketability of farmed products has not been exploited to any extent.

Genetic selection is the standard approach used in animal breeding programs to improve marketability. Development of a particular commercially desirable trait through genetic selection will depend on what characteristic is being considered and the genetic variation around the trait. Rate of progress will depend on gene frequency, the number of loci involved, and the degree of linkage. If the trait is rare, selection may be very slow. Selection programs, therefore, will benefit from having sufficient numbers of trout with which to begin. The limits of response to selection will be determined by the gene content of the base population. At least 100 effective mating pairs is recommended as a brood stock population, but periodic out-crossing would further increase the probability of discovering new genes to enhance the process.

Desirable traits are more apt to show up in certain families rather than appearing randomly throughout the brood stock. Therefore, selective breeding programs have to safeguard against the predominance of certain family members representing the next generation of brood stock. When that occurs, the effective mating population becomes much smaller, and the chances for inbreeding and fixation significantly increase. Fixation occurs when the two genes at any locus are identical. If fixation involves deleterious genes, the fitness of the population is reduced. Therefore, in addition to increasing the gene pool, a large population would also decrease chances of inbreeding depression.

High growth rate is one of the desirable traits commonly selected for in commercial brood stock, and significant gains in selection programs on growth rate have been achieved. Selection limits exist for any trait and occur when all favorable alleles for that trait have been brought to fixation. Response to selection occurs rapidly at first and then tapers off as fixation among the alleles progresses. Cost effectiveness in selection programs for improved growth rate, therefore, may be limited to five or ten generations. Even at that rate, however, brood stock selection programs require a considerable time commitment.

Molecular biology is becoming a frontier in the agriculture industries, and alteration of a genome can be nearly immediate by applying available technology. Transfer of copies of genes into the vertebrate genome was initiated in the mouse.[206] Isolated genes injected into the pronucleus of the mouse zygote integrated with the genome and underwent normal mitotic growth. Gene transfer was later demonstrated in rabbits, sheep, pigs, and goldfish.[207,208] Genetic attributes of fish have a good potential of being changed to meet specific objectives of producers.[209] Trout strains engineered for certain commercially desirable traits would provide immediate changes in brood stock development and could render standard selective breeding programs obsolete. The scope of possibilities for improving rainbow trout for commercial use through molecular biology extends well beyond the most optimistic expectations of any selective breeding program.

C. ADULT HOLDING FACILITIES

Adult holding facilities for rainbow trout brood stock and returning steelhead are often raceways set aside for such purposes. In some instances, holding containers may be a wooden picket trap built to intercept migrating fish at a diversion weir. Summer-run steelhead were routinely caught and held for ripening in such facilities on the Dungeness River on the Olympic Peninsula. Positioned in the home river and protected from high velocities, traps provided good flushing flows to irrigate the confined fish until they were ready to spawn, without causing noticeable stress.

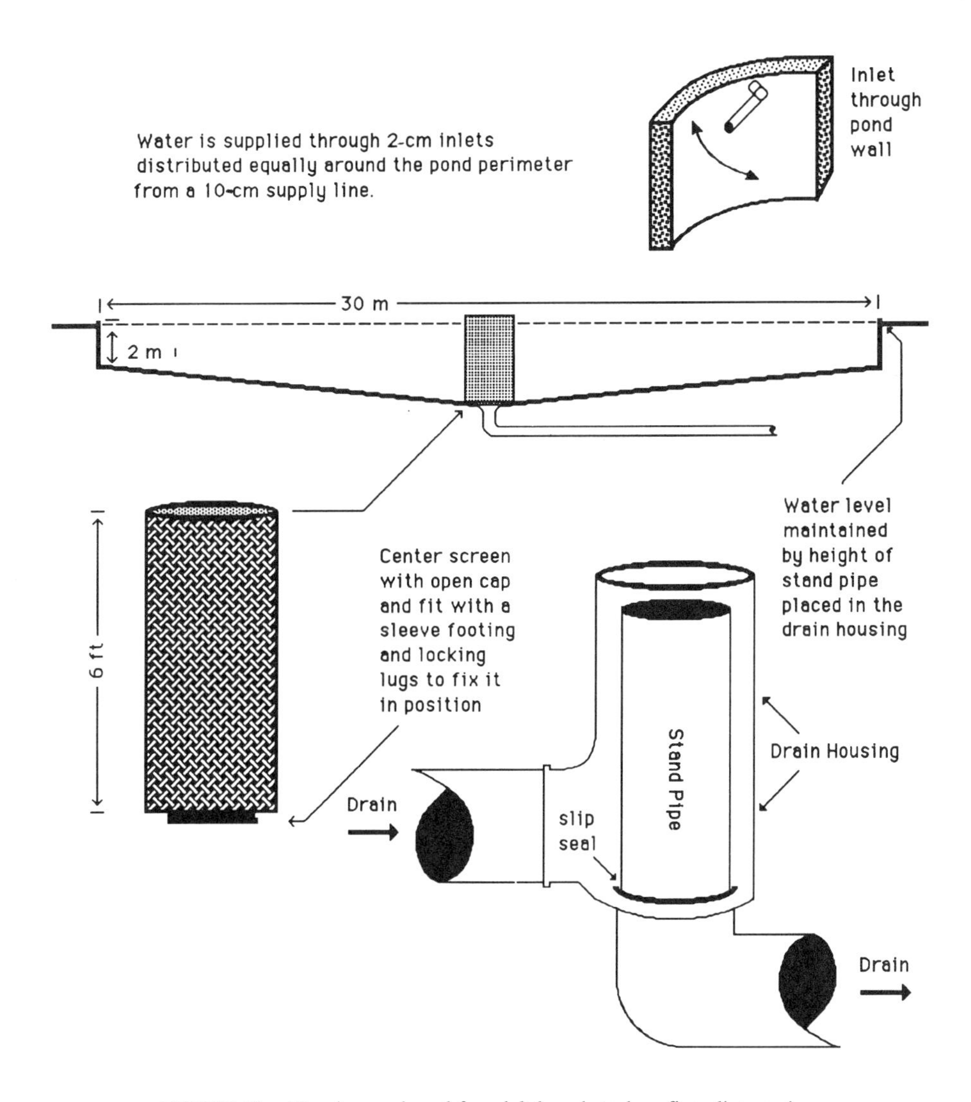

FIGURE 10. Circular pond used for adult brood stock or fingerling rearing.

When steelhead return within days of becoming ripe for spawning, such holding containers may be sufficient. Raceways and traps, however, do not provide the best adult environment for good fish health in conjunction with long-term holding. Brood stock holding facilities need not be designed with rearing pond constraints, but should have sufficient space to allow the fish to retreat from stress and provide freedom of movement. Large circular ponds of 20 to 40 m diameter, 3 to 4 m deep, with a stable or fixed substrate floor sloping to a center drain provide good holding conditions for domesticated or wild adults of any size (Figure 10). Water entering the pond should enter through several perimeter subsurface inlets to eliminate injuries from jumping, and to create a sustained velocity to encourage continuous swimming. The salmon and steelhead return pond at the University of Washington is a good example of such a facility. The pond is a gravel-lined basin 33 m in diameter and 2 m deep at the center with a flow of 43 l/s inducing a perimeter velocity of 30 to 60 cm/s. Up to 2,000 adult salmon have been held in the pond without showing gamete inviability problems from stress.

To reduce handling and eliminate the uncertainty of when fish are ready to spawn, arterial inflowing conduits provide channels that ripe fish will enter when seeking spawning sites. Salmonids have evolved a spawning repertoire that includes migrating upstream to spawn in areas with good irrigation velocities. Even domesticated fish maintain this strong positively rheotactic behavior at maturation, and inflow channels entering the holding pond will attract fish that are seeking spawning sites. Rather than seining and handling all of the brood stock several times a season in search of ripe fish, those ready for spawning isolate themselves. Handling in these situations is much reduced and often limited to just artificial spawning.

D. ARTIFICIAL SPAWNING

Since rainbow trout will spawn more than once, live spawning is usually practiced. Readiness to spawn is determined by the appearance of the fish and by gently testing for loose gametes. Males are generally ready to spawn when the first females ripen, but in domesticated stocks, synchrony is not guaranteed. At maturation both males and females change from their normal bright coloration to darker shades, and some to dark brownish green. Both sexes develop the pronounced characteristic red band along the lateral line. Males will show the elongated lower jaw which hooks up slightly. Females will develop an extended abdomen, accommodating the developing ova.

Females will demonstrate readiness to spawn by the abdomen changing from a firm to a very soft texture. A ripe female lifted from the water in a horizontal position will indicate her readiness to spawn by the sagging abdomen and loose eggs. The genital papilla will be slightly swollen and extended, and often reddish in color. A light pressure exerted on the abdomen and squeezing an inch forward of the papilla of a ripe female will express eggs from the urogenital pore. This is the final test to demonstrate spawning readiness. The paired ovaries will ripen nearly simultaneously, and all of the eggs can be freed from the skeins and be ready for fertilization at about the same time.

Lifting the male from the water and exerting pressure in the same way as with the female will express white milt if he is ripe. The testes do not ripen at once, however, but over several days, which allows the males to spawn over a period of time with several different females. In hatchery practice, the males can be used several times over the period of ripening, since the sperm at the terminal end of the testes are mature first, and they progressively ripen to the anterior end. Care should be taken not to express too much milt when testing because with domestic stock, a small amount may be all that is available at any one time.

Spawning can take place after anesthetizing the fish to reduce stress during struggling and to reduce the possibility of injury to the gametes. Anesthetic is placed in an aerated, water-filled tank for treating the fish. Tricaine methane sulfonate (MS-222), and quinaldine or a combination thereof (40 ppm MS-222 + 10 ppm quinaldine) are some of the more commonly used anesthetics. Anesthetized fish will roll over when relaxed enough to handle, and they should be removed shortly thereafter for spawning. Placing excessive numbers of fish in the anesthetic tank should be avoided. Anesthetic dose levels should not exceed concentrations that roll the fish over in less than 30 seconds.

When spawning is started, the female must be wiped dry of all water that might reach the eggs. Water on the eggs will initiate water absorption through the egg capsule (water hardening) and prevent the eggs from being fertilized. The female should be lifted from the water by the caudal peduncle with her head kept slightly down from the horizontal position at all times until the eggs are to be expressed. At that time, grasp the caudel peduncle in one hand while cradling the body, dorsal side up, in the other arm across your midsection with the tail end slightly down. Exert pressure on the female's anterior abdomen by cradling the abdomen in the palm of your hand. Exert the pressure toward the posterior abdomen, squeezing with the thumb and fingers. Don't squeeze any more than necessary, and don't massage or strip back and forth on the abdomen until the free-flowing stream of eggs has stopped, and then only between the ventral fins and genital papilla. Unnecessary massaging can injure the female, break eggs, and spread albumen over the good eggs, which can substantially reduce subsequent fertilization success. If glassy (hard) eggs are observed in any number, the egg lot should be discarded. The female was over-ripe, and the eggs will not be viable.

Live spawning most often leaves eggs still in the skeins of the female, especially if the brood fish have been held in ponds where lack of exercise does not afford enough activity to loosen the eggs from the skeins, or if the female was not quite ready to spawn. Steelhead will generally have sufficient ovarian fluid to flush all but a few of the eggs from the body cavity. Domesticated rainbow trout will have to be spawned a second time, 48 hours after the first spawning, to remove the remaining eggs. In some instances it is necessary to spawn a third time. In those situations, only the eggs that freely flow should be removed during the first spawning, to reduce the stress from handling. At the second spawning the ratio of ovarian fluid to egg mass will be much higher and eggs can be expressed more easily.

Spawning by injecting normal saline through the urogential pore will remove the eggs with much less stripping and often eliminate the need for the second spawning. A plastic bulb syringe adapted with a tube to a saline reservoir is all that is necessary After the free-flowing eggs have been expressed into the egg container, the syringe tip is inserted through the urogenital pore and sufficient saline is pumped into the body cavity to extend the abdomen to its previous volume. The abdomen can then be massaged with the tail up to dislodge the eggs from the skeins and viscera, and then with the tail tilted down, gentle pressure is exerted on the abdomen again. With this method, the eggs are usually expressed into a colander to strain the eggs from the ovarian fluid and saline. This routine can be repeated a second time if it is apparent that some eggs were retained. If eggs are kept at ambient temperature and away from moisture and sunlight, they can be held 48 hours without reduced viability. In the presence of water, eggs will remain susceptible to fertilization for only about 60 seconds, but after 30 seconds the fertilization rate drops rapidly.

For fertilization, milt can be expressed directly on the eggs from a male by holding him as you would the female and squeezing between the ventral fins and genital papilla. If you need to conserve the milt supply, males can be spawned into a separate container and their milt pooled for distribution to the eggs of several females. In these situations it is necessary to wipe water from the male to prevent sperm activation that will result from contact of milt with water. Once water activation occurs, motility of the sperm will last about 20 seconds. Fertility drops after 25 seconds. If milt is kept cool (not on ice) and away from sunlight, it can be held up to 24 hours without markedly reduced viability. Only a few drops of milt will be sufficient to fertilize an egg lot, but distribution of the milt through the egg mass is necessary to assure fertilization.

E. SEX ALTERATION

The commercial trout industry is dominated by the pan-size product and has been very successful in developing the portion-size market. Growing larger fish for market has been inhibited somewhat by early maturation of the males. Rapid growth rate tends to accelerate maturation, and many of the rapidly growing males will mature sexually at less than one year of age. Secondary sex characteristics associated with maturation, such as weight loss, and change in body conformation and pigmentation markedly reduce the market value of the fish. Sex alteration to all female production stock or sterilization of males can provide producers with improved ability to grow large size rainbow trout without precocious maturation.

Various researchers have achieved sex alteration through hormone administration during development or early feeding. The most successful treatment in sterilizing rainbow trout has been accomplished with methyltestosterone.[210] Effective treatment is dosage and time dependent. Administering the hormone to developing embryos in an immersion bath and subsequently feeding them as fry with hormone-treated food has been highly successful. Both males and females can show a high degree of sterilization, with no subsequent development of secondary sex characteristics. Rainbow trout sterilized with methyltestosterone at the University of Washington were kept to an age of eight years, which was five years past the death of the controls. Although growth rate during the first year was slightly less than that of the untreated controls, there was no growth cessation in the hormone-treated fish at the time of maturation among the controls when growth rate normally declines. The largest of the treated fish eventually grew to well over 13 kg.

Other studies have reported some cranial and body deformation in conjunction with the use of methyltestosterone on salmon. Rainbow trout body conformation appeared normal, however. The methodology is simple, inexpensive, and special breeding stock isn't required. The level of hormone administered is minute, it is rapidly cleared from the body of the fish, and the fish are very small at the time of treatment. Treatment details for sterilization are as follows:

1. The immersion bath — 17 α-methyltestosterone

Solution concentration	300 μg/l
Number of immersions	4 times
Stage of development treated	255, 315, 375, and 435 degree days
Duration of immersion	One hour

2. Feeding treatment — 17 α-methyltestosterone

Concentration	20 mg/kg of food
Application	Sprayed on food
Length of treatment	1st 90 days of feeding
Amount of feed fed	3% body wt/day

Applying heat-shock to recently fertilized eggs has also been effective in sterilizing rainbow trout. Immersion of eggs shortly after fertilization in a 35°C water bath for 1 minute (or 27 to 30°C for up to 10 minutes) will interfere with the extrusion of the second polar body and induce triploidy that results in sterility in both sexes. Survival can be very high, but the triploids do not grow quite as fast as diploid controls. The problem with triploid rainbow trout is that, while they are sterile, the males still show secondary sex characteristics, which obviates the purpose for sterilization.

Thorgaard et al.[211] indicated that heat-shock treatment after fertilization with radiation-inactivated sperm can be used to produce gynogenetic diploid females, which can then be sex-reversed as phenotypic males having XX chromosomes. Methyltestosterone is used to masculinize the fry when fed in low doses (1 to 3 mg/kg of food). The gynogenetic step is necessary to assure an all-female production stock. Feeding methyltestosterone without first having the gynogenetic step will produce phenotypic males, but they will have both XX and XY chromosomes. While this process is somewhat complicated, it is effective in eliminating precocious maturation problems with males. Simply using gynogenetic diploid females as production stock would be undesirable because any recessive deleterious genes would be homozygous. As a source of XX sperm, however, they are quite satisfactory.

F. CARE OF THE EGGS

Once milt is placed on the eggs, water is added to activate the sperm and initiate water hardening. The eggs are then gently washed with repeated rinses of water to remove excess milt. Any organic debris will promote the growth of fungus (*Saprolegnia* sp.). If the washing process is not immediately followed by placing the eggs in incubators, they should be held in a container with water at three times the volume of the egg mass for 30 minutes to water harden. During the water-hardening process, their high sensitivity to shock can result in increased mortality if the eggs are handled.

Water hardening is the process whereby the water entering through the microvilli of the egg capsule fills the space between the vitelline membrane and the capsule by osmotic and hydrophilic pressures.[212] The soft eggs spawned from the female are made turgid by water absorption, and the supportive sphere that encompasses each egg will enhance early incubation. During the first part of water hardening, the eggs are adhesive. If the egg surface touches another surface, suction through the microvilli will hold the surfaces together until water hardening is 90% complete. Complete water hardening takes several hours.

Antiviral iodophor is added to the water-hardening solution in many cases to help reduce the vertical transmission of infectious hematopoietic necrosis (IHN) virus in commercial rainbow trout populations. The treatment method is 30 minutes exposure to the iodophor (buffered Wescodyne or Argentyne) at a concentration of 150 ppm.

It is important to keep the eggs out of sunlight. If exposed to sunlight for only a few minutes, newly fertilized or developing eggs will suffer mortality. Lightproof covers are recommended for hatchery incubators and even exposure to artificial light or indirect sunlight should be kept to a minimum.

Sensitivity of the developing eggs to any physical shock exists from 48 hours post-fertilization to 9 days (95 degree days) of incubation at 10.5 °C. Sensitivity starts before the blastula is formed and continues through epiboly until blastopore closure. Up to that time the vitelline membrane covering the embryo and yolk can easily rupture and mortality will result. At blastopore closure, embryonic cell layers have engulfed the yolk and substantially increase resistance to physical disturbance. Picking and careful handling can take place at that time without mortality. Hatchery personnel generally wait until the eyes are pigmented (15 days or 157 degree days) before handling eggs, because that stage is easily identifiable and one is assured that sensitivity to handling is no longer a problem.

If eggs are going to be transferred or placed in deep incubators away from access, they have to be in a condition that requires little care. Incubation mortality ranges from 5% to 35% of the eggs spawned. Much of that loss is infertile or undeveloped eggs that appear as "blanks". The vitelline membrane of blank eggs will eventually break down, which causes the yolk material to coagulate. Once the integrity of the membrane is compromised, fungus will invade, and it most often quickly spreads to engulf and smother adjacent viable eggs. Therefore, where there will be limited access, the process is enhanced by physically shocking the eggs to identify the blanks before continuing incubation. This process is applied at the eyed stage when sensitivity to handling has passed. Shocking methods vary, but often just siphoning or pouring the eggs from one container to another will provide a sufficient jolt for the process. Coagulated yolk turns white and the dead eggs readily show up in the egg mass. Devices to remove eggs range from specialized tongs with narrow wire loops used to grasp the eggs and pipettes with a syringe bulb to a continuous siphon tube with a stop check to control the rate of suction.

When large numbers of dead eggs have to be removed at once, the salt floatation method can work quite well since eggs with translucent yolk and coagulated yolk have a slight difference in density. Dead eggs will float when the salt solution is at the proper concentration, and the dead eggs will thus be separated from live eggs. Care has to be taken in preparing the solution because the effective range is quite narrow A salt solution of approximately 10.3% will separate trout eggs, but the precise salt concentration should be determined through experimentation with each egg batch. Once the salt solution is working, the concentration will have to be adjusted periodically because dead eggs will absorb salt. Starch has been used with success, also, and starch is not absorbed by dead eggs. The container holding the salt or starch solution should be large enough to accommodate the incubation compartment or tray. As the tray is submerged, the dead eggs float to the top and can then be skimmed off with a dipnet.

Eyed eggs are quite hardy, but are vulnerable to low dissolved oxygen. Oxygen levels that fall below saturation will result in some reduction in development rate. If oxygen levels drop to 3 mg/l, emergence timing can be delayed by as much as 3 weeks. Oxygen levels below 2 mg/l will result in suffocation because of the competition for oxygen among the incubating eggs. Dissolved oxygen concentrations below 7 mg/l should be avoided. If mortalities are occurring from suffocation, they will be indicated by the presence of embryos that turn white, and immediate action is called for to prevent further losses.

If flow problems occur, incubator baskets or trays should be taken out of the water and kept moist by sprinkling water over the egg compartments. Sufficient oxygen will be provided by that method because the wetted egg surface will enhance oxygen exchange until the eggs can be returned to water. Eggs can be maintained in this manner until hatching. Exposure to incubation water experiencing gas supersaturation is not a problem prior to hatching because the egg capsule is a limiting osmotic barrier, and the respiratory needs of the embyro maintain a large enough oxygen differential between the internal and external environments that excess gas can be accommodated.

It is important to maintain good records on egg inventory throughout incubation. Counting or measuring eggs has been made much easier since electronic balances are now widely available. Total weight or volume measurements can be taken, and then sample weights or volumes counted to enumerate the egg population. Use of a perforated measuring cup is probably the quickest method. The volume of eggs contained in the cup is determined and a subsample is counted to correct for egg size. The cup is filled with eggs and the volume can be easily converted to egg number.

G. INCUBATION AND INCUBATORS

Eggs and alevins are incubated in trays, baskets, boxes, cylinders, or on various types of substrates in numbers from a few hundred to a million at a time. The important principle is that eggs and alevins need sufficient irrigating velocity to replenish the oxygen and flush away metabolites. If the flow requirement is satisfied, nearly any container can be used for incubation. Conservation hatcheries often incubate trout in trays containing up to 10,000 eggs, and stacked in units of 8 or 16 to reuse the water cascading down from tray to tray in the stack (Figure 11). The trays are in the form of a screened frame to hold the eggs, which are further confined by a screen cover and placed horizontally in a compartment. Water upwells through the egg mass in the tray, overflows the compartment, and is then guided through a channel to fall to the next compartment beneath. Flows through incubator trays are from 8 to 30 l/min.

Standard hatchery egg baskets (15 × 28 × 61 cm) with elongated screen mesh are still used in some hatcheries. The baskets sit in 5-m-long troughs, 30 cm wide, and 20 cm deep (Figure 12), and are suspended on hangers to keep them above the floor of the trough. Water flows in one end of the trough and out the other, irrigating the eggs. Baffle plates direct the flow from underneath the basket to the surface of each successive compartment.

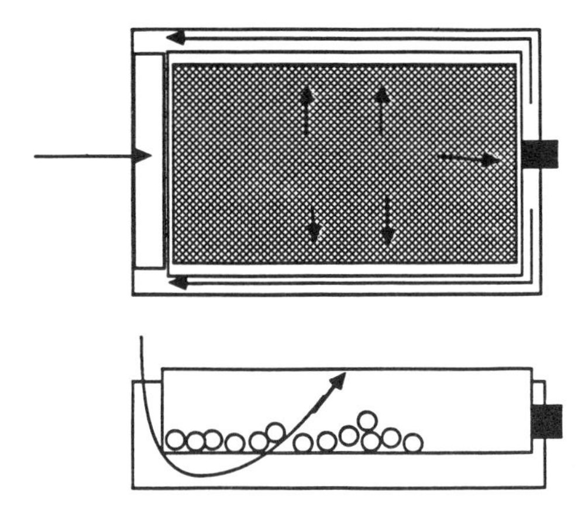

FIGURE 11. Tray incubator showing upwelling flow pattern.

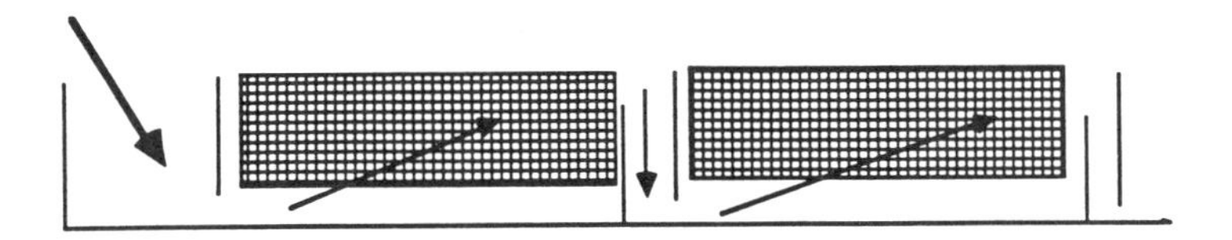

FIGURE 12. Trough and basket showing flow pattern.

As the eggs hatch, the alevins drop though the elongated screen mesh of the basket to the trough floor, leaving unhatched eggs behind. Six to eight compartments are contained in each trough, and flows range from 23 to 46 l/min. The tighter the baskets fit in the compartments, the better the eggs are irrigated and the less water is required.

If egg lots are to be kept isolated for any reason, they can be incubated in small buckets that have a screen colander inserted to hold the eggs and alevins off the bottom, and extending above the water line to keep eggs from being carried away in the overflow. Water is supplied by a tube inserted through the side of the bucket near its base. Exiting water overflows the rim of the bucket.

Commercial incubators for rainbow trout usually accommodate large numbers of eggs. A common incubator for trout eggs in commercial hatcheries is a deep incubation cylinder containing up to 100,000 eggs. A horizontal screen keeps the eggs off the bottom of the container where the water enters, and allows the irrigating flow to upwell throughout the egg mass and overflow from a spout on top. Depending on temperature, for the first 10 days or so of incubation the upwelling velocity is set low enough to keep the eggs from moving, because of their sensitivity to any disturbance. After 10 days, when the eggs can tolerate movement, upwelling flows should then be increased so as to slightly lift and slowly roll the egg mass. Circulating the eggs in this manner assures good oxygen distribution, and helps keep any fungus present on dead eggs from engulfing live ones.

H. CARE OF THE ALEVINS

Quality of fry emerging from the incubator is strongly determined by the incubation environment. Alevins have a fixed yolk store that supplies energy for maintenance, growth, and exercise. At any given temperature, yolk absorption after hatching occurs at a constant

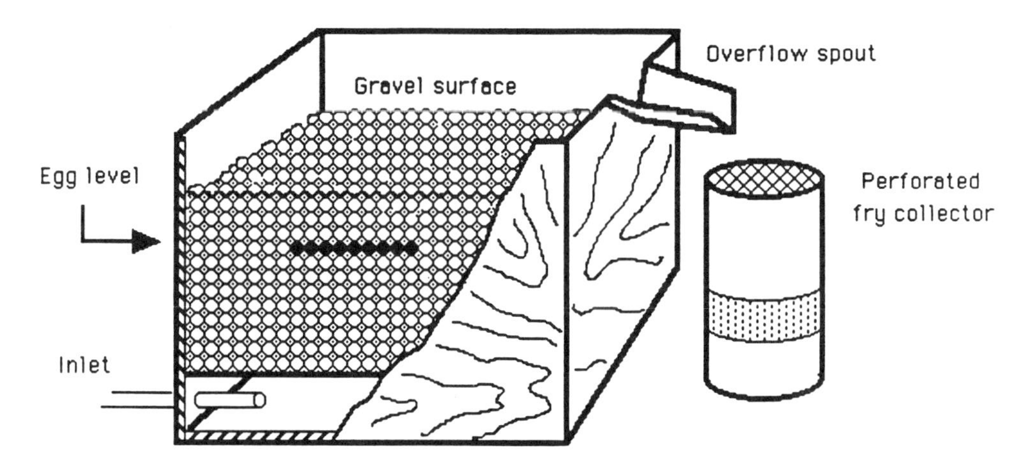

FIGURE 13. Upwelling gravel box incubator.

rate.[213] Maintenance and exercise take priority over growth, and if the alevins are forced to exercise, growth will be sacrificed. In situations where alevin movement is restricted such as in the presence of substrate or in low velocities, yolk conversion efficiency to alevin weight gain can be higher than 62%. However, with high levels of activity, yolk conversion to alevin weight may be as low as 35%, and the resulting fry size at complete yolk absorption only half of what it might otherwise have been. Fry size at emergence or initiation of feeding is very important. Larger fry will start feeding better and perform better in ponding. If fry releases are made into streams, larger fry will be stronger, more competitive, and less vulnerable to predation.

Substrate is often employed in incubators to limit the activity of alevins once hatching takes place. Typical substrates are artificial turf, bio-rings, short plastic tubes, or washed and sized stream gravels. Substrates are added to incubators at the eyed stage after shocking to remove the dead eggs. In some instances the freshly fertilized eggs are placed on the surface of the substrate, and the alevins bury themselves after hatching.

Streamside incubation boxes are sometimes used in supplemental enhancement programs. Their size varies from 0.03 m^3 up to nearly 6 m^3, and the design varies from gravel-filled boxes with screen floors to perforated pipes used to deliver incubation water, to boxes with only 5 to 10 cm of gravel placed directly on the floor where they function similar to hatchery troughs. A typical box will have a false screen floor to hold the substrate above the water inlet for good irrigating flow distribution, 30 cm of washed gravel, and a spout for the exit flow at the top (Figure 13). Gravel must be sized to correspond to the size of the alevins and not provide such large void spaces that the young fish slip through and incubate on a densely crowded, two-dimensional floor. Eggs can be spread on the gravel surface as long as the box is covered to exclude light. This will allow the eggs to be picked if necessary and make the operation much less labor intensive. Once hatching takes place the alevins will bury themselves.

Another benefit of substrate incubation is that emergence, and thus initiation of feeding, is at the volition of the fry. In the case of rainbow trout, emergence and initiation of feeding occur very near the time of complete yolk absorption. Even among individual alevins from within a single egg lot, complete yolk absorption will range over a 2 week period. If released at the same time, some will be ready to feed much earlier than others. Volitional emergence allows the fry more appropriate timing to initiate feeding, and growth variation or pin-head losses will not be so apt to occur.

If unfed fry plants are to be made, it is best from the standpoint of behavioral considerations if the fish are distributed in their stream immediately upon emergence. Emergence occurs at night and by first light of dawn the fry should be distributed away from their incubation site. This is the best reason to have supplementary hatchery operations take place on the native stream. Incubation temperatures will correspond to the proper timing of emergence, and if emergence is volitional, the fry will enter the stream when they are ready to start the hiding and feeding phase of stream residence.

When rainbow trout fry have been incubated without substrate in a hatchery before release, they are conditioned to a situation different from what they will experience in their native environment. Hiding and escape behavior do not develop the same as in the stream, and moving shadows signal something other than hazards. Swim-up in a hatchery trough or open incubator will occur much earlier than swim-up from gravel incubation, with as much as 25% of the yolk stores remaining. Alevins planted too early have very low survival rates. Swimming is difficult, the yolk sac makes them obvious, and they are not ready to exercise escape and evasion maneuvers.

Light also affects the activity level of incubating alevins. Shortly after hatching, alevins become photonegative and seek areas out of contact with light. If alevins are denied darkened retreats, they will continue their searching activity and lose energy from exercise rather than putting it toward growth. Incubation systems should employ covers to keep alevins in near total darkness.

Avoiding low oxygen is important to good alevin health, but the flow passing through the incubator is usually sufficient for alevin needs. Gas supersaturation is a problem, however, and alevins are highly susceptible to levels of dissolved nitrogen only slightly above saturation. Chronic exposure to only 102% nitrogen saturation will result in gas-bubble disease with bubbles forming underneath the yolk membrane of alevins. This situation often occurs when alevins are brought out from deep incubators into shallow troughs where the change in hydrostatic head allows dissolved gas to come out of solution.

Water velocity during alevin incubation should not be excessive, or energy losses from the fish trying to maintain orientation will result in the most severe loss of growth. The old, deep troughs used in conservation hatcheries that employed stacked trays for hatching and alevin incubation had flows that often exceeded 80 l/min. That level of flow passing through alevin masses on each tray created velocities that continuously swirled the alevins in circles. Such an abnormal incubation environment for fish that are normally in the protection of the gravel substrate not only wasted their energy stores, but subjected them to extreme exercise weeks before they were developed sufficiently to handle that stress.

V. REARING OF FRY AND FINGERLINGS

The science of rearing fish is a matter of applying the appropriate methods to fish as any good farmer applies his knowledge in growing farm animals or crops. There is a significant amount of art involved, which is best defined as knowing your fish and watching for the signs that first forecast potential problems. Adequate coverage of all the information associated with fish rearing is not be covered in this chapter; the reader is referred to either Klontz[214] or Piper et al.[45] for more exhaustive coverage of the topic.

Rearing situations take place under either intensive or extensive conditions. The difference in the two methods is simply whether or not the fish are fed with artificial feed. Although this chapter concentrates on rainbow trout culture under intensive conditions, extensive trout culture offers the culturist various options that are also mentioned.

A. EXTENSIVE CULTURE

Under extensive culture operations, trout feed only on the natural food they capture. This situation places a limit on the biomass of trout that can be grown per hectare of pond area, with the result that many of the problems experienced under intensive rearing conditions never become an issue in extensive pond culture. Oxygen levels, carrying capacity, and disease potential are generally not problems. Photosynthesis provides sufficient oxygen even if the pond is static, and the amount of space available for the fish alleviates the disease concerns as long as temperatures don't get too high. In extensive rearing ponds, 60 kg of rainbow trout produced per hectare represents the upper end of the production curve. Even with that level of yield, however, it would not be very profitable unless a farmer operated a fee fishing business where total weight of production wasn't necessarily the objective or measure of success.

Water depth is important in extensive pond culture. Deep basin retreat areas are necessary for refuge from high temperatures and predators. A basin in excess of 3 m is desirable, and should preferably be deeper to provide thermal stratification for retreat during hot weather and a reservoir of 4°C water in the winter. Spring water entering the pond will alter the design criteria, depending on the temperature and oxygen level of the spring source.

Shallow areas in the pond should be deep enough to limit access to the fish by wading birds. Blue herons will swallow 0.5-kg trout and spear those too large to eat. If the water is at least 1 meter deep in the shallow areas, such birds are confined to the shore and their predation on the fish will be greatly reduced. Even mallard ducks are serious predators on small fry. Gulls can be the most destructive bird predator on fingerling and market size fish.

Stocking strategies will be to plant fry that are feeding well from a nursery pond that has been prepared with a good culture of natural food. Springtime productivity of copepods and cladocera are ideal feed for small trout. Once feeding has been well initiated and growth is occurring, fry can be stocked at 5 kg/ha. If 10- or 20-g fingerlings are stocked, 750 to 1,000 fish per hectare should be released, based on the expected recovery of 0.5-kg fish after 1 year. Predation by birds and other fish will account for 90% of the mortality during the first months of growth.

Fee fishing ponds are stocked only with catchable fish, and the density will be determined by factors other than the rearing potential of the pond. As fish are caught, they are replaced from an intensive rearing pond close by. A lake or reservoir several hectares in size can be stocked for rearing and fishing. In those instances, 20-g fingerlings can be planted at 5,000 fish per hectare, and fishing will begin to thin them out as they reach 100 g apiece.

B. INTENSIVE CULTURE

Intensive culture refers to the situation where fish depend on artificial feed. If natural food is available, it will have little influence on the weight gain of the fish because fish densities are too high to benefit from the quantity of natural food available. Although most of the problems experienced by commercial fish farms are very different from those responsible for poor success of conservation hatcheries, in many instances the problems have to do with the same basic rearing technology. Intensive culture of rainbow trout needs to be given much more attention in both conservation and commercial culture situations. The following sections address some of the subjects that concern new fish farmers.

1. Pond Design

Raceways are the most popular rearing chambers (Figure 14) in the United States. Raceways are long rectangular ponds in which the water enters at one end and exits at the opposite end. Raceways are good ponds if one has access to relatively high flows. The trend

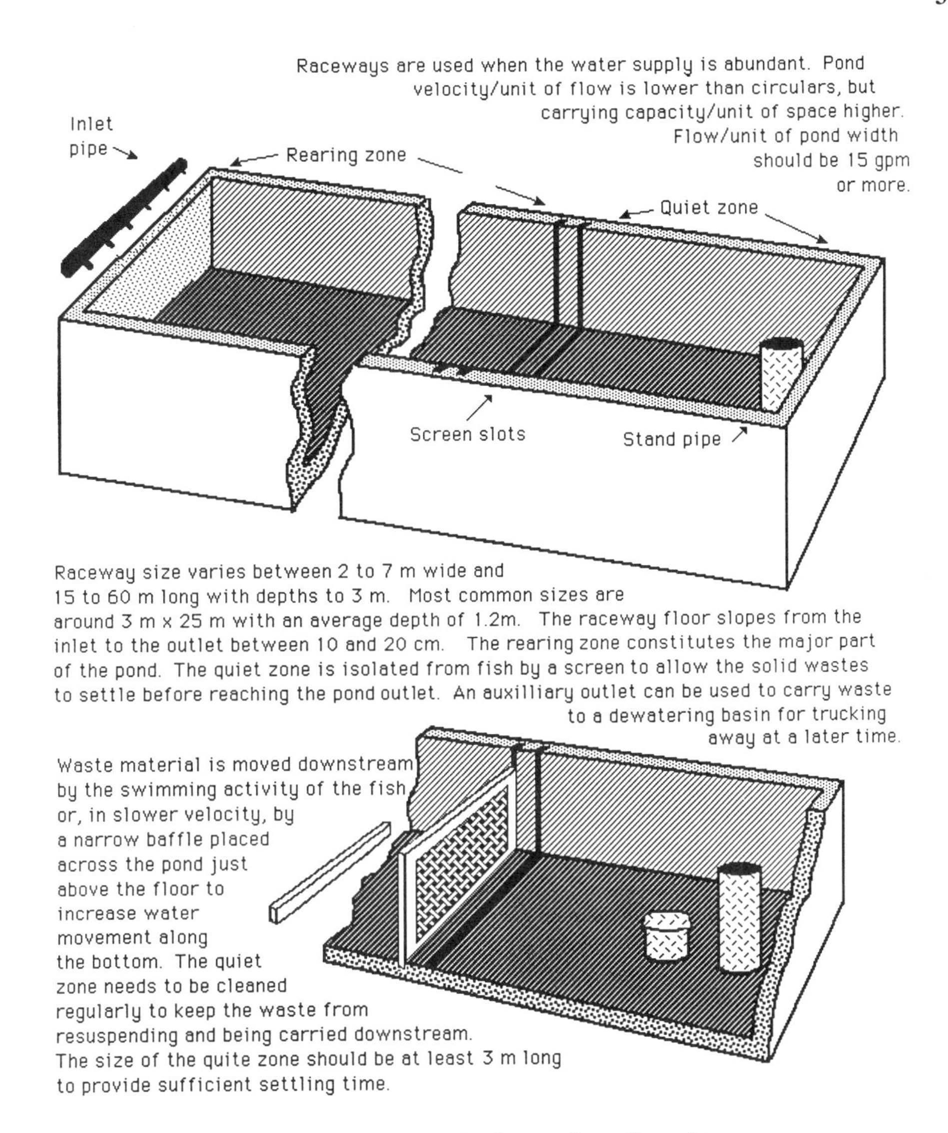

FIGURE 14. Raceway and quiet zone for settling solids.

to raceways was set by conservation hatcheries with relatively unlimited water supplies available, and the high flows kept the ponds well flushed. The advantages that raceways offer, however, fade if water has to be used to maximize efficiency. The inlet end of the raceways will have maximum oxygen and the lowest levels of metabolic waste materials. As the water passes down the raceway, the rearing environment degenerates with metabolic materials increasing and oxygen decreasing. Moreover, the velocity in a raceway is at a minimum for good fish exercise.

To provide good flushing velocities, the cross-sectional area of raceways can be designed to accommodate whatever flow is available. The best design of a raceway system is to operate ponds in a series, with discharge flow from one running into the next in sequence. Such a stair-step system should not be loaded much higher than the same volume would be loaded in a single pond because of ammonia build-up, but it will allow better aeration by plunging water between the divisions. Plunge height or fall (Figure 15) between raceways

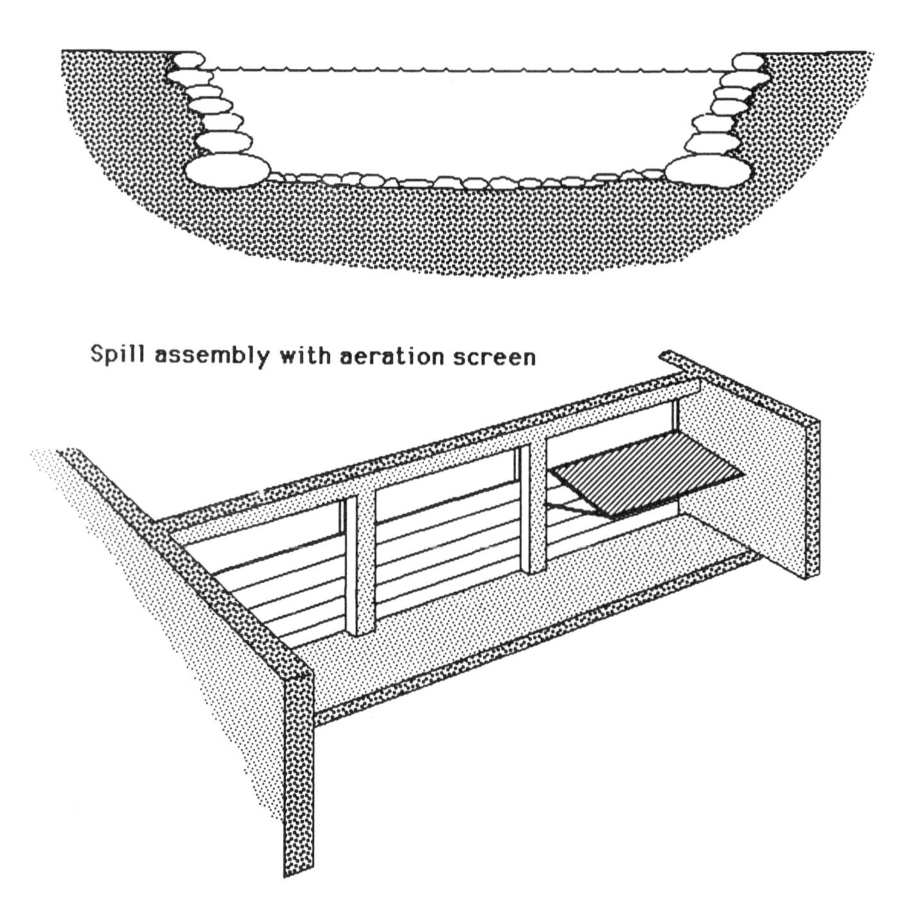

FIGURE 15. Aeration drop between ponds.

should be greater than 0.6 m, and water should splash on a broken surface or fall through a screen before entering the next level to maximize the air-water interface for exchange. Some ammonia will be exhausted from such a system, but the major reconditioning process is oxygen recharge and, to a lesser extent, carbon dioxide elimination. Screened quiet zones upstream from the outlet will permit solids to settle from the water column before the water flows to the next pond.

Circular ponds are becoming popular again because they are good rearing containers when water supplies are limited (Figure 10). The outlet is located in the center and the inlet water can be distributed across the radius of the surface. Inflow jets directed at an angle from 30 to 60°C will adjust the velocity independent of the flow passing through the pond and allow more rapid elimination of the waste material. The higher velocities in circular ponds improve exercise conditions for the fish. Since inflowing water can be delivered across the radius of the pond, in addition to having excellent circulation conditions, the dissolved oxygen levels will be more uniform than in linear raceways. The circular flow in the pond allows self-cleaning. A cowling around the outlet screen will vacuum the waste off the floor and provide a quiet zone for it to settle in the center or discharge over the stand pipe.

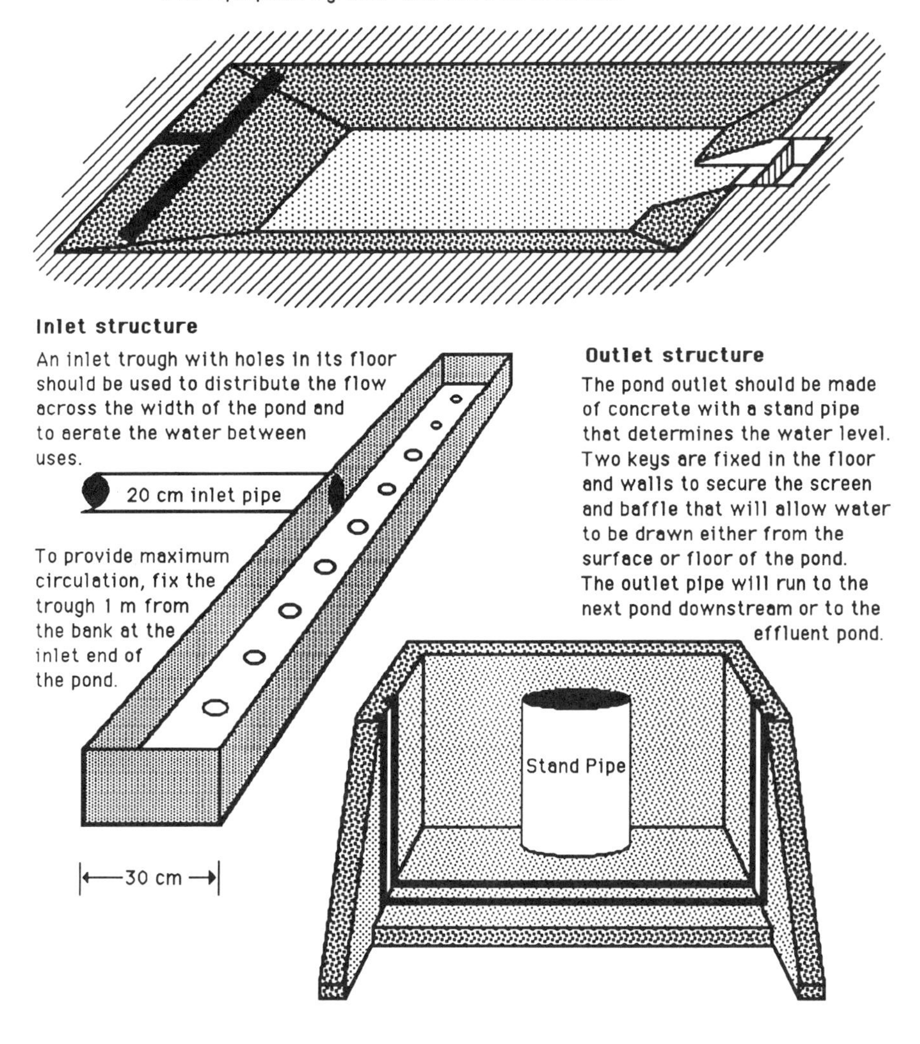

FIGURE 16. Earthen pond used for rearing.

Concrete ponds are the most costly, but the easiest to maintain. Raceway size varies, but can range from 3 × 10 m to 7 × 33 m, and from 0.6 to 2 m deep, sloping 15 to 30 cm over their length. Circular ponds will range in diameter from 1 to 1.5 m for small fish, from 2 to 7 m for fingerlings, and from 10 to 30 m for larger fish. Depths vary from 0.6 to 2 m, and the floors slope to the center to assist in draining and cleaning.

Earthen ponds (Figure 16) are still used in both commercial and conservation fish farms and are the most common ponds used for trout in Europe. Water velocity rates through earthen ponds are usually low because the ponds are generally larger than hard-walled facilities. Earthen ponds are difficult to keep clean, and removing the fish can be a problem. A hardened or concrete moat may be placed in the outlet area to collect the fish when the pond is drained. Earthen ponds can be very good fish rearing facilities, but loading levels should be conservative.

2. Ponding and Fingerling Care

Initial rearing containers for newly emerged fry can be troughs or small ponds. Good circulation and flushing rates are the important components of the fry rearing phase. Flows should not create excessive exercise, but sufficient velocity should be present to keep the fish adequately exercised for physical conditioning before release. About one fish length per second is an appropriate velocity for trout fry, and the same criteria can be followed throughout the rearing period as the fish grow. It has been demonstrated for Atlantic salmon that exercise for 2 to 12 months before release resulted in twice as many fish returning to the hatchery than was shown for unexercised controls.[215] Even in commercial hatchery programs, exercise can be beneficial in improving food conversion, growth, and muscle characteristics.[216]

Preconditioning fish to avoid predators, or to avoid moving images above the water surface, would assist survival success after release from ponds. Tests with salmon demonstrated that preconditioning fingerlings to predator silhouettes with an electrical impulse assisted in making the young salmon avoid predators after release.[217] Similar tests on steelhead have not been undertaken, but in conservation hatcheries, preconditioning the fingerlings before release should be considered as routine practice.

Prevention is the best disease treatment that one can have in a hatchery. Assuming that one obtains disease-free stock, a good rearing environment is the key to minimize problems.[218] Good care means that sufficient oxygen and pond flushing rates have to be provided and maintained. Although trout will survive at oxygen levels below 3 ppm, the minimum concentration for trout culture is often suggested as 5 ppm, which is the minimum standard for waste effluent required by water resource agencies. However, the minimum oxygen necessary to provide rainbow trout with optimum growth and good health is 7 ppm. The rearing objective, therefore, should be that oxygen levels not drop below 7 ppm to maintain optimum conditions for the fish. To maintain that standard, aeration would have to be provided throughout the pond system.

Maintaining the oxygen level with surface aeration using atmospheric air will not exceed saturation (Table 3). Therefore, aeration at the surface with atmospheric air is safe, but less efficient than other systems. Atmospheric air has the N_2:O_2 ratio of 4:1, or four times as much nitrogen as oxygen. Under hydrostatic head, the level of gas at saturation will increase proportionally to the increased atmospheric pressure and may then become a problem when the water rises to shallower depths and the dissolved nitrogen comes out of solution. Dissolved nitrogen will cause gas-bubble disease at even low levels of supersaturation.[219] However, when oxygen is being used and the water is less than saturated, as is characteristic of most intensive rearing ponds, supersaturated dissolved nitrogen will not be a problem as long as the total dissolved gas concentration doesn't exceed saturation.

The oxygen requirement problem is not just the need to recharge the water after each successive use. Even with fish held at optimum density, the oxygen level in the upper end of the raceway is much higher than the concentration near the outlet. As oxygen concentrations are reduced progressively down the length of the raceway, reduction in feed efficiency and growth rate will be experienced by fish residing primarily in the lower half of the pond. Recharge methods, therefore, have to address maintaining satisfactory oxygen levels in the entire raceway at all times. This will require aeration or oxygen injection down the length of the pond.

Water flow entering the pond should provide flushing rates that replace the volume twice an hour. Flow distribution has to be uniform as well, to assure that no waste or dissolved solids are concentrated. The design of the outlet screen, standpipe, or overflow structure is important to assure that water circulates efficiently. Raceways pose the greatest problem in

TABLE 3
Dissolved Oxygen (mg/l) at 100% Saturation at Sea Level

Temp (°C)	Elevation (m) 0	300	600	900	1200
0	14.6	14.1	13.6	13.1	12.6
1	14.2	13.7	13.2	12.8	12.3
2	13.8	13.3	12.9	12.4	12.0
3	13.5	13.0	12.5	12.1	11.7
4	13.1	12.6	12.2	11.8	11.3
5	12.8	12.3	11.9	11.5	11.1
6	12.4	12.0	11.6	11.2	10.8
7	12.1	11.7	11.3	10.9	10.5
8	11.8	11.4	11.0	10.6	10.3
9	11.6	11.4	10.8	10.4	10.0
10	11.3	10.9	10.5	10.1	9.8
11	11.0	10.6	10.3	9.9	9.5
12	10.8	10.4	10.0	9.7	9.3
13	10.5	10.2	9.8	9.4	9.1
14	10.3	9.9	9.6	9.2	8.9
15	10.1	9.7	9.4	9.0	8.7
16	9.9	9.5	9.2	8.8	8.5
17	9.7	9.3	9.0	8.7	8.4
18	9.5	9.1	8.8	8.5	8.2
19	9.4	8.9	8.6	8.3	8.0
20	9.2	8.8	8.4	8.1	7.8

this regard because of their low operating velocities. If water is drawn from near the pond floor at the outlet end of the pond, at least the dead flow spaces that often occur around the outlet can be avoided.

Waste material should be removed from the pond as often as possible. Solids will settle out of the water column onto the pond floor and become resuspended by activity of the fish. As this happens, fine material will be resuspended and become an irritant to the gills of the fish. Heavier material will settle out quickly, but if it is left in the pond the wastes will begin to disintegrate into finer material that will resuspend and add to the debris irrigated across the gills. Baffle boards placed across the pond 2 to 3 cm off the floor at intervals down its length (Figure 14) will increase the velocity of water passing under the baffles and help sweep the material to the outlet.

3. Feeding Strategy

Once complete yolk absorption approaches, feed should be made available to the small fish to allow early feeders access to food. In any single egg lot, individual alevins will vary in completing yolk absorption by at least 2 weeks. If fry are ponded at what is considered complete yolk absorption, some will have been without food for several days and will have begun to starve, resulting in pin-heads and mortality. Swim-up, or that point when alevins leave the trough floor and enter the water column, signals when food should first be administered. It is best to crowd newly feeding fry to initiate rapid conversion to an artificial diet. Usually a fish density of 15 kg/m^3 of water will suffice for about a week before spreading them out to lower densities of 5 kg/m^3. Thereafter density will depend on the objective of the hatchery program.

Feeding rates vary for small fish depending on temperature and fish size (Table 4).

TABLE 4
General Feeding Rates in Percent Body Weight for Different Sizes of Rainbow Trout Held in Water of Various Temperatures[44]

Temp (°C)	Body length (cm)						
	<2	2.5	5.0	7.5	10.0	15.0	25
3	2.7	2.2	1.7	1.3	0.9	0.7	0.5
4	3.0	2.5	2.2	1.7	1.1	0.8	0.6
5	3.3	2.8	2.2	1.8	1.3	0.9	0.6
6	3.6	3.0	2.5	1.9	1.3	1.0	0.7
7	4.0	3.3	2.7	2.1	1.5	1.1	0.8
8	4.1	3.4	2.8	2.2	1.6	1.2	0.8
9	4.5	3.8	3.0	2.4	1.7	1.3	0.9
10	5.2	4.3	3.4	2.7	1.9	1.4	1.0
11	5.4	4.5	3.6	2.8	1.9	1.5	1.0
12	5.8	4.9	3.9	3.0	2.1	1.6	1.1
13	6.1	5.1	4.2	3.2	2.2	1.6	1.1
14	6.7	5.5	4.5	3.5	2.4	1.8	1.2
15	7.3	6.0	5.0	3.7	2.6	1.9	1.3

Feeding frequency should be 12 to 16 times per day initially, taking care to only dust the surface and not create debris in the water column. Bacterial gill disease will be encouraged with excess food in the water. It is important to follow the old rule of "feed the fish and not the water", to make sure the food is not falling to the trough floor. Although fry can pick food off the bottom, if it remains there for just a short time it absorbs water, loses nutrients, and disintegrates when disturbed.

Feeding frequency and percentage of body weight (BW) fed daily decrease as the trout grow. After the first month, feeding frequency can drop to eight times a day, until the fry are converted to demand feeders. Table 4 provides a general guide to feeding rate. Individual circumstances, including desired market size and time at which that size is to be reached, will determine the actual feeding level selected. As temperature increases to 15°C, feeding rate increases to the optimum level for growth. At higher temperatures optimum feeding rate will level off and then decrease as temperatures increase further.

It is useful to determine condition factor and food conversion rates to evaluate the feeding program. Wasted feed is wasted dollars and unnecessary costs need to be avoided. The condition factor (C_f) is a relationship between length and weight of the fish, and is expressed as:

$$C_f = \frac{100 \times \text{weight in g}}{\text{length}^3 \text{ in cm}}$$

Young trout just beginning to feed will have a condition factor that is around 0.95. Alevins with large yolk sacs will have condition factors higher than 2.0. Condition factor drops as yolk is absorbed and length increases. It is helpful to make feed available to fry when their condition factor is approaching 1.0, which is a better criterion than swim-up because the amount of light available without substrate present will alter the timing of swim-up. If the condition factor falls below 0.90, the fry are beginning to starve and muscle weight is being used for energy.

As trout begin putting on weight, the condition factor will increase to 1.0 again, and with more body bulk it may increase to 1.2 or 1.3. As a rule, condition factors above 1.3

are a sign that too much fat is being laid down and food conversion efficiency, which is the weight gain obtained from the amount of food offered, is dropping. Some strains of trout are very bulky for their length and may normally have a condition factor of 1.3 to 1.4. Knowing what is expected from the strain of fish being reared will provide the criteria that should be established. A great deal of adaptation will be required to fine-tune the feeding procedure on each farm.

Watching food conversion efficiency will also alert the culturist to potential problems with the feeding procedure. Very young fry will convert food to fish flesh at a ratio of better than 1.0. Often, conversion ratios can be as good 0.85 if the diet is well balanced in nutrients. As the fish grow, food conversion ratios will increase and may reach 1.2 or 1.5. Conversion ratios for young fish will generally not be lower than 1.0, and larger fish should not have conversion ratios higher than 1.5. If ratios go outside of the desirable range, too much food is being fed and some is being wasted. It was not uncommon to see food conversion ratios of 2.2 when automatic feeders first came on the scene. Since feed costs can amount to 60% of a farm's production expense, the importance of conversion efficiency is often the difference between profit and loss.

As Table 4 demonstrates, temperature is a major factor in determining how much to feed. Warmer temperatures will increase the amount to be fed by two to three times that recommended at lower temperatures. At cold temperatures (<3°C), more than 96 hours are required to evacuate the gut after feeding. Growth potential at those temperatures is very low, so only maintenance rations are fed, and as infrequently as once or twice a week. When fry are fed at maintenance levels, it is important to feed the fish to satiation to assure that all of them receive food. If fish are fed more frequently, but given the same daily maintenance ration, the aggressive fish will take the food, leaving the smaller fish with no feed and increasing the range in size.

Food particle size is important to assure that feed is not too large or too fine for the fish being fed. Fry are often started on mash or fish meal and then crumbles, before starting on small pellets. The expanded esophagus of a trout is about 1/25th of the fish's length. The pellet size recommended is approximately 1/50th of fish length, so a trout 7.5 cm long will be fed pellets 1.5 mm (1/16 inch) in diameter at the narrowest dimension. Feed should be put through a sieve to remove fines before feeding to reduce waste and degradation of pond water quality. The fines can be repelleted with a binder such as cooked processing waste and fed as a moist feed to larger fish.

The manner in which feed is delivered can be a major factor in how much is wasted. Hand feeding, automatic or timed feeders, and demand feeders are the three methods used. Hand feeding is time consuming, but control can be maintained and feed can be placed where the fish are with minimum waste. One has to watch that this activity does not become routine and spread feed over the water without determining the location of the fish. The best means of determining problems with fish health is to watch how the fish respond to hand feeding. Fish will generally begin to go off feed hours before any problem is apparent, and high losses can be averted with good responsive care.

Automatic feeders can be placed within reach of most of the fish, and feeding can be programmed at predetermined rates to occur at any time interval, day and night. Automatic feeders have the potential disadvantage of higher food wastage. If fish go off feed as a result of activity by predatory birds or other problems, distribution of feed continues regardless. The advantage of automatic feeders is that growth rates can be effectively planned to meet production or market projections without having to hold back inventory.

The most popular feeding technique used by commercial hatcheries is demand feeders, where the fish feed at will by activating a mechanism that releases feed. Except with carefully planned automatic feeding programs, demand feeders have shown as good or better food conversion efficiencies as other feeding schemes. When maximum feeding rates are desired, demand feeders work effectively. When less than a maximum ration is desired, however, demand feeders are filled accordingly, and efficiency can be reduced. Feeding regimes on limited rations need to be carefully controlled to prevent size variability and reduced food conversion.

Fin erosion, characteristic of hatchery fish, is affected by feeding practices. Fish densities higher than 5 kg/m^3 of water can be demonstrated to induce dorsal fin erosion. Less than maximum ration and lower feeding frequencies have also been shown to increase fin erosion.

4. Grading

If the fry are to be reared to fingerlings or to the smolt stage, feeding routines are important to develop with the size at release kept in mind. Hatchery fish are easily overfed and can grow too fast. Grading the fish will help maintain uniform size classes and improve food conversion efficiency. Small or more timid feeders remain small when competing with aggressive, faster growing fish. If the pond population is graded and kept in size groups within 25% of their mean, performance within the size groups will be significantly enhanced. Unless grading is practiced, the rationale behind the feeding strategy of optimum ration size is lost. Larger fish will consume nearly all the feed, and small fish will be left with little or no feed.

Grading fish is a routine that most hatcheries follow, but the philosophy between conservation and commercial hatcheries is different. Conservation hatcheries grade to maximize feed efficiency and growth benefits among all genotypes represented and to reduce the number of fry dropping out of the population. Commercial hatcheries often grade as a means of culling to rid themselves of the less efficient feeders that can become a feeding liability. Some Norwegian salmon farmers will cull in excess of 50% of their population. The cost of starting with twice as many eggs even at five cents apiece becomes insignificant when keeping inefficient feeders amounts to a 30 to 50% increase in production costs.

Various graders are employed in hatcheries. Perhaps the best is the in-pond grader that does not require handling of the fish. In-pond graders are framed vertical bar grates, with the vertical bars spaced equidistant at intervals determined by the size of fish to be separated. The frame is positioned between wing barriers that are placed across the pond after the fish have been crowded downstream. When the grader is in place, fish sufficiently small in size to pass through the bars will swim upstream and separate themselves from the fish retained by the grader. A sequence of grading frames can be placed in the pond to allow separation of the fish into as many size intervals as desired.

Box graders are one of the most common types found in hatcheries. A box grader is literally a box with the floor made of spaced bars through which the smaller fish can swim. The grading bars can be PVC pipe or metal rods fixed at equal intervals across the floor, and supported with cross braces so they remain rigid. The box needs to be deep enough to prevent the fish from jumping over the sides and should provide enough space for the fish to maneuver. Fish immediately sound to the floor when frightened, so when they are placed in the box with a dip net, the smaller fish will slip through the bottom and the larger will be retained by the grading bars. Lifting and sinking the box will encourage any remaining smaller fish to pass through. The fish left represent the size range selected. Further separation can be achieved by employing different grading frames in the box. Grader efficiency is better if the conditions in the grader remain the same and the fish are kept in water at all times. Size separation will be different if the fish are allowed to lay against the bars out of water, or if the mass of fish is too dense. Handling induces stress in salmonids.[220] Avoiding excess handling or abusing the fish by keeping them out of water will reduce stress and enhance the ability of the fish to recover from grading.

Large commercial graders are available that use the same principle, but they generally require the fish to fall across wetted surfaces rather than remaining in water. With the exception of commercial graders used prior to harvest, grading fish out of water should be avoided.

5. Carrying Capacity in Conservation Hatcheries

Conservation hatcheries may rear rainbow trout to a release size of about 100 g and often much smaller. Physical characteristics of the fish, such as fin definition, broad tails, and low condition factors are important for performance when the fish are released. Pond loadings, therefore, should be much lower than what may be appropriate for commercial hatcheries.

From the standpoint of rearing conditions, crowding will cause behavioral stress in rainbow trout, which is normally a nongregarious species, before problems occur from low dissolved oxygen or high ammonia concentrations. Stress within the rearing environment, or any reduction in fish health when hatchery fish are released, will have serious survival implications. Low rearing densities with good flushing rates will provide a good rearing environment for rainbow trout, and fish cultured in such conditions are expected to survive better after release than fish raised in higher densities.

The maximum density that can be carried in a pond is determined by careful calculations of the amount of oxygen required to metabolize feed at the ration level fed, the temperature of the water, and the ammonia generated. However, for reasons mentioned above, the calculated maximum density should never be used in conservation hatcheries. The criteria recommended for conservation hatcheries, therefore, represent the set of guidelines adapted from information developed by Piper et al.[45] for U.S. Fish and Wildlife Service hatcheries. The guide differentiates between Density Index and Flow Index, based on pond volume and flow rates. Both factors have to be considered and the limiting factor determined by which one is reached first.

Density Index is the carrying capacity based on the weight of fish per unit of pond volume. The index addresses the behavioral needs of trout for space. The standard recommended here for conservation hatchery fish is lower than that used by Piper et al.[45] and adjusts loading density on fish length in the equation:

$$2 \times \text{length in cm} = \text{weight in kg/m}^3$$

where 2 is the conversion constant. Densities determined for fish of different sizes, therefore, will increase as fish increase in size. Density of fish 15 cm in length will be 30 kg/m^3 with this equation. Since the density index is not normally the first to be exceeded, this criterion is liberal and does not become restrictive unless the ponds are small for the loading level permitted by the flow available.

The Flow Index is the carrying capacity based on the weight of fish per unit of flow. The "F" factor in the equation developed by Piper et al.[45] is based on the amount of oxygen available for life support and growth. The index compensates for percentage of body weight fed as the fish increase in length. An "F" value is determined for each temperature and elevation, assuming oxygen saturation, and reflects the relationship of weight of fish per unit of flow with fish size as follows:

$$F = \frac{W}{l \times I}$$

TABLE 5
Flow Index Values Recommended for Rainbow Trout in Conservation Hatcheries, Compensated for Temperature and Altitude

Temp	Elevation (m)					
(°C)	0	300	600	900	1200	1500
5	.090	.083	.076	.070	.064	.059
6	.083	.076	.070	.064	.059	.055
7	.076	.070	.064	.059	.055	.050
8	.070	.064	.059	.055	.050	.046
9	.064	.059	.055	.050	.046	.042
10	.059	.055	.050	.046	.042	.039
11	.055	.050	.046	.042	.039	.036
12	.050	.046	.042	.039	.036	.033
13	.046	.042	.039	.036	.033	.030
14	.042	.039	.036	.033	.030	.028
15	.039	.036	.033	.030	.028	.026
16	.036	.033	.030	.028	.026	.024
17	.033	.030	.028	.026	.024	.022
18	.030	.028	.026	.024	.022	.020
19	.028	.026	.024	.022	.020	.018
20	.026	.024	.022	.020	.018	.017

where W = weight of fish in kg, L = length of fish in cm, and I = flow rate in l/min. The "F" values given in Table 5 are lower than those presented by Piper et al.[45] The values are conservative and are based on the densities recommended to provide more healthy fish in conservation hatcheries. With this equation a rearing unit at 7°C and 300 m elevation, containing 15-cm trout, and receiving an inflow of 1,100 l/min, will have a carrying capacity of:

$$W = .070 \times 15 \times 1100 = 1155 \text{ kg}$$

If the rearing unit is a pond 3 m wide, 1 m deep, and 15 m long, it will have 45 m^3 of volume. Using the 1,155 kg carrying capacity determined by the flow index, 45 m^3 of volume would result in a fish density of 25.7 kg/m^3, which is below the 30 kg/m^3 determined as the maximum loading permitted by the density index. Pond space and available flow will determine which index limits the loading level.

6. Carrying Capacity in Commercial Hatcheries

The efficiency of intensive culture systems in commercial hatcheries is related to the quality of the rearing environment in a manner similar to conservation hatcheries. However, the behavioral concerns are not an issue in commercial hatcheries, and carrying capacity can be determined simply by calculating the effect of loading on oxygen and ammonia levels present. Overloading the carrying capacity of a commercial farm pond will also result in reduced success in terms of poor fish health and reduced feed efficiency; so care has to be given to address both oxygen available for use and the ammonia build-up.

Among the several methods used to determine carrying capacity, the equation used here[219] relates just to the food being fed and not the size or number of fish involved. A given amount of food will require a given amount of oxygen for metabolism, and it will produce a given amount of ammonia. The quantities of oxygen and ammonia involved, therefore, will determine how much feed can be fed in a given flow. Based on available water flow, a given daily ration can be fed that will not exceed the rearing limits for oxygen and ammonia. That quantity will then determine the carrying capacity in fish biomass as a function of maximum food to be fed over percent body weight fed.

The oxygen required and ammonia produced will depend on food quality, but for purposes of illustration it is assumed that for every kilogram of food fed, 0.25 kg of oxygen is required, and 0.032 kg of ammonia is produced.

The model uses the relationship:

$$N = (0.25)/(0.00143 \times Ox)$$

and:

$$p = R/N$$

where

N = l/min required/kg of food fed

0.25 = kg O_2 to metabolize 1 kg of food

0.00143 = conversion constant

Ox = inlet oxygen minus outflow oxygen (Oa − Ob)

p = kg of food fed

R = total rate of flow in l/min

When Oa is 9 ppm and Ob must meet effluent standards of 5 ppm, then 4 ppm oxygen is available for rearing, and for every kg of food fed, 43.7 l/min flow is required as shown below:

$$N = \frac{0.25}{4 \times 0.00143} = 43.7 \text{ l/min/kg}$$

With a total flow of 1200 l/min, total food fed per day based on oxygen is

$$p = \frac{1200 \text{ l/min}}{43.7 \text{ l/min/kg}} = 27.5 \text{ kg of feed}$$

The size or number of fish in this situation are variables determined simply by the maximum weight of food the oxygen level allows to be fed. If 27.5 kg of feed will be fed fish to be hauled to farm ponds when they reach 7.5 cm in length, capacity is calculated for 7.5-cm fish just before hauling. If the temperature is 10°C, 7.5 cm fish will be fed at a rate of 2.7% of BW, which will allow 1,019 kg of biomass (27.5 kg/.027 = 1,019 kg) or approximately 212,000 4.8-g fish (C_f 1.06) to be accommodated in the hatchery pond before hauling without exceeding the oxygen requirements. However, if the feed is for fish to be harvested at 25 cm from a pond at 15°C, just before harvest those fish will require a feeding

TABLE 6
Percent NH_3 of Total Ammonia Relationship to pH and Temperature

Temp (°C)	pH 6.0	6.5	7.0	7.5	8.0	8.5
4	.01	.03	.12	.37	1.10	3.39
8	.02	.05	.16	.50	1.58	4.82
12	.02	.07	.21	.68	2.12	6.40
16	.03	.09	.29	.92	2.86	8.52
20	.04	.13	.40	1.24	3.83	11.18

rate of 1.3% BW, which will represent a biomass of 2,115 kg, or a population of 11,280 fish (C_f 1.2).

Carrying capacity is determined by the amount of oxygen required to metabolize the weight of food fed, and hence is limited by the environmental oxygen supply. Carrying capacity will be affected by a different feed with a different oxygen requirement, but the greatest influence will be altering environmental oxygen. As readily seen, oxygen recharge can elevate Oa and increase the carrying capacity proportionally. If oxygen is no longer limiting because of recharge capability, the ammonia level becomes the limiting parameter for carrying capacity.

The same approach[219] is useful in estimating the ammonia build-up from feed metabolism. Total ammonia (NH_3 + NH_4^+) generated from feed is composed of ionic (NH_4^+) and molecular (NH_3) forms. NH_3 is toxic to trout at about 0.02 ppm. NH_3 declines logarithmically with pH and therefore becomes a hundredfold less of a problem as pH decreases from 8.5 to 6.5 (Table 6). Carrying capacity based on ammonia is calculated to determine the flow needed to keep NH_3 below the 0.02 ppm toxic threshold. The equation is the same with additional variables involved. Total ammonia generated per kg of food fed is estimated at 0.032 kg. However, in this case pH and temperature determine what percent of the total ammonia generated is in the toxic form, and that value (Table 6) is used to determine carrying capacity.

The model uses the relationship:

$$N = (0.032 \times r)/(0.00143 \times 0.02)$$

and:

$$p = R/N$$

where

N = l/min required/kg of food fed

0.032 = kg NH_3 + NH_4^+ produced/kg of food fed

r = % NH_3 of total ammonia present (Table 6)

0.00143 = conversion constant

0.02 = ppm max NH_3

p = kg of food fed

R = total rate of flow in l/min

Given a temperature of 12°C and a pH of 8.0, based on ammonia the model will predict that for every kg of food fed, 23.7 l/min flow is required as shown below:

$$N = \frac{0.032 \times 0.0212}{0.02 \times 0.00143} = 23.7 \text{ l/min/kg}$$

With a total flow of 1,200 l/min, total food fed per day based on ammonia is:

$$p = \frac{1{,}200 \text{ l/min}}{23.7 \text{ l/min/kg}} = 50.6 \text{ kg of feed}$$

Similarly, the carrying capacity determined by the ammonia level will be the maximum weight of food that NH_3 allows to be fed. If 7.5 cm fish at 12°C are to have a feeding rate of 3% BW, then 50.6 kg of feed will support 1,687 kg of fish in a pond. However, if the pH was 8.5, the minimum flow required to maintain NH_3 at a safe level would be 71.6 l/min/kg of feed, allowing only 16.8 kg of food fed daily, and a loading level limited to 560 kg of fish. Without biofiltration of ammonia, successive reuse of pond water through a serial raceway system will rapidly reach ammonia limitation.

7. Fish Releases

Conservation hatcheries release resident trout into streams or lakes targeted for providing an improved sport fishery. Numbers released, therefore, depend on the management objective, but if planting is meant to initiate a self-sustaining resident population, numbers cannot exceed the productivity of the system or its ability to feed the population.

The same is not true for anadromous steelhead plants into streams for mitigation or enhancement purposes. The river in these cases is used simply as a home stream transportation system to get the smolts to sea. Any need to feed during migration often does not enter the equation, especially when the fish may be in route less than a few hours. The problem with this situation is that hatchery fish do not undergo the physiological changes involved in smoltification at one time. Smoltification is size related and year-old steelhead may not smolt until they reach a certain size threshold, which in some populations is around 13 cm. If large numbers are released, and 20% or more are not ready to smolt, or have passed the point of smoltification, the process will have planted a large population of temporary resident fish. The capacity of the stream may be exceeded by such an operation, with the resident rainbow trout and residing wild juvenile steelhead having increased competition from the hatchery fish. Resident fish may not be displaced by hatchery fish, but poor survival of wild fish may result because of the extra strain on food resources.

This issue launched considerable research into developing smolt indices that would predict when fish should be released. The problem, however, is that such indices could not provide the answer except in broad terms. Up-river fish in large river systems will start migratory behavior at a different physiological state than coastal populations. Moreover, fish readiness to migrate is an individual response, and population indices only indicate when most individuals are ready. The key to successful timing of release to maximize smolt success is to let the fish select their own migratory timing. Volitional releases require more time and special facilities, but the question of whether or not the fish are ready to migrate is resolved.

Steelhead usually migrate at 2 years of age. Hatchery steelhead reach migratory size in less than a year because of enhanced growth from artificial diets. The result of hatchery releases of 1 year-old smolts is that most of the adults return a year earlier than wild fish,

TABLE 7
Desirable Water Quality Characteristics for Trout Culture

Parameter	Desirable range
Dissolved oxygen	7.0 ppm to saturation (best level)
	5.0—7.0 ppm (limited growth)
pH	6.7—8.5
Alkalinity	80—200 mg/l as $CaCO_3$
Carbon dioxide	<2.0 mg/l
Calcium	>50 mg/l desirable (4—160 mg/l)
Zinc	<0.04 mg/l at pH 7.6
Copper	<0.006 mg/l in soft water
	<0.3 mg/l in hard water
Iron	<1.0 mg/l
Ammonia (NH_3)	<0.02 mg/l constant
	<0.05 mg/l intermittent
Nitrite (NO_2^-)	<0.5 mg/l
Nitrate	0—3 ppm
Nitrogen	<100% total saturation
	<110% total gas pressure at saturation
Suspended solids	<80 mg/l
Dissolved solids	20—500 mg/l
Hydrogen sulfide	<0.002 mg/l (should be below detection limits)

as 3-year fish. The loss of the additional year at sea results in a return size of only 2 to 4 kg, compared to 4-year wild fish returning at 6 to 8 kg. The best way to improve the size of hatchery steelhead, therefore, is to slow freshwater growth and release smolts at age 2.

Improved hatchery fish size is not related just to changing hatchery management procedures, however. Sportsmen dedicated to preserving the native stock of steelhead in the Solduck River entering the Straits of Juan de Fuca on the Olympic Peninsula of Washington worked with the Washington State Department of Game (now Washington Department of Wildlife) to catch wild Solduck River brood stock for spawning, and to avoid having the Chambers Creek hatchery stock planted in the stream. Sportsmen caught the steelhead with sports gear and transferred the fish to a Washington Department of Game hatchery for spawning. The progeny were marked for identification and returned to the Solduck River for release as smolts. The success of such an effort to preserve a wild run was shown to be possible by the return of marked steelhead of the 8 kg class, even though they were raised and released under routine hatchery practices. Steelhead return size was demonstrated to be influenced not just by culture strategy, but to have a genetic component involved in size at return as well.

VI. WATER SOURCES AND EFFLUENT

A. WATER QUALITY

Water quality (Table 7) and quantity are the first concerns one must have about the physical site being selected for a fish culture facility. Water volume has to be adequate to provide the rearing flows necessary, and it is best to have spring water as one source of supply at the station to provide flexibility for temperature control. Conservation hatcheries often use river water where sufficient quantities can be diverted. Commercial operations have different rearing objectives that don't have to fit into ambient temperature and timing schedules. Minimum flow requirements, however, for both types of facilities should be around 300 l/sec, although many commercial farm ponds operate on much less.

Three sources of water are available to fish farmers. Spring water or ground water are the sources that often seem to be the most desirable because of the constant temperature they provide. The supply is usually insufficient for commercial operations, however. A second source is lake water, which gives several good options if the lake is large and deep, but such sites are rare. The third is surface water, the most abundant, but generally considered the least desirable because weather conditions have such a strong influence on quality. Historically, spring water was sought for hatchery operations and surface water was avoided because of the potential problems it brought. However, surface water offers the most promise for future development and should be given a second look. Quality of surface water is more than adequate for salmonid culture; a fact readily apparent when one looks at where most of the freshwater habitat for trout exists.

The advantages and disadvantages of each source are given in the table:

Ground Water

Advantages	Disadvantages
1. Temperature is constant	1. Temperatures are often too cool for good rearing
2. Temperature is higher than ambient winter air temperatures	2. Often deficient in oxygen, but with nitrogen supersaturation
3. Source is generally free of silt	3. Volume is usually limiting
	4. Pumps are often necessary
	5. Mineral levels can be high

Lake Water

Advantages	Disadvantages
1. Temperature control option	1. High organic silt
2. Heat sink	2. Low winter temperature
3. Abundant supply	3. Summer gas supersaturation can occur
	4. Pumps are necessary

Surface Water

Advantage	Disadvantages
1. Abundant supply	1. Organic silt high
2. Pumps not always required	2. Stream debris
	3. Prone to floods
	4. Ambient water temperature
	5. Winter ice problems

B. EFFLUENT TREATMENT AND WATER REUSE

Effluent from fish farms is viewed as industrial waste by the U.S. Environmental Protection Agency. Fish waste, however, is not the typical industrial effluent that one expects from the use of the term. Fish waste is a nutrient, and if it were applied to nutrient-poor streams it would provide an excellent boost to stream productivity and fish production. Effluent becomes an issue when it exceeds the capacity of the environment to accommodate it, or when the luxuriant growth produced from the nutrient clogs waterways.

The first requirement is to remove solid wastes. Settling of solid waste in raceways or auxiliary ponds is presently practiced by trout farmers and it will concentrate most of the solids generated in the pond. Screened quiet zones upstream from the outlet greatly enhance the settling process. A screen 3 or more meters from the raceway outlet, or 1.5 m from the center drain of a large circular raceway, will provide a quiet area and keep the fish from disturbing the settled solids. Wastes will accumulate in the quite zone for easy removal with a siphon or pump. Baffle boards placed across the pond, 2 or 3 cm off the floor at intervals down the length of the pond (Figure 14), will increase the velocity of water passing under the baffles and help sweep the material to the quiet zone.

The second problem has to do with dissolved nutrients. Biofiltration of the dissolved solids is practiced by some trout farmers to improve water for successive reuse. The major problem associated with that technology is the magnitude of water required for treatment and the nature of the waste. Unlike domestic sewage, which is a low volume of concentrated waste, fish-rearing effluent involves relatively large flows of very dilute waste. Typical systems used in municipal sewage treatment facilities, therefore, are not readily applicable to fish farms. New systems developed for fish farm effluent treatment include biofiltration through wetland application of wastewater.

Water reuse is practiced to increase production per unit of water available to the farmer. This task must initially address the oxygen supply which is the first limiting parameter of a rearing system. Various aeration devices or oxygen recharge systems exist, but generally the application of aeration technology does not require a high cost unless liquid oxygen is employed. As covered in the section on pond design, falls between raceways may suffice for aeration needs. Trout farms routinely produce pan-size trout at 1.8 kg/l of flow in a single pass system, and with aeration it can rise to 3 kg/l.

Where terrain does not permit drops between ponds, other aeration systems will be necessary. Aeration by mechanical agitation increases the dissolved oxygen more quickly than natural surface exchange because such action provides a greater air-water interface. Maintaining the oxygen level by surface aeration is safe and will not exceed saturation, but may be less efficient than other systems. Total amount of dissolved oxygen available will depend on temperature and pressure.

Increasing dissolved oxygen with methods other than surface aeration, such as air or oxygen injection, will require greater cost. The amount of a gas dissolved in water is determined by the partial pressure and solubility of the gas and is independent of other gases. Air has the N_2:O_2 ratio of 4:1, or four times more nitrogen than oxygen, so aeration provides only 20% of the oxygen that would be available compared with using pure oxygen. Therefore, when saturated with pure oxygen in the absence of nitrogen, water will carry five times more oxygen than at air saturation. The capacity of water to hold oxygen in solution, however, is greatly affected by hydrostatic pressure, and will increase by approximately 10% over atmospheric pressure with each meter of depth increase. Water oxygenated under 3 m of head can absorb 30% more oxygen than at the surface. At 15°C and 600 m elevation, 9.4 ppm oxygen is at saturation on the surface, but at 3 m of depth the water would hold about 12.2 ppm. Therefore, injecting oxygen at depth in pond wells (between ponds) will provide a mechanism to increase the oxygen level more efficiently than attempting oxygenation at atmospheric pressure. Delivery systems for pure oxygen must include an exact metering device for delivery at the appropriate point in the system to take advantage of hydrostatic pressure and prevent waste.

If air is injected under hydrostatic head, nitrogen will follow the same gas laws as oxygen. Dissolved nitrogen may become a problem when water recharged at depth rises to the surface and gas begins to come out of solution. Nitrogen will cause gas-bubble disease when fish are exposed even at low levels of chronic supersaturation. However, when oxygen is less than saturated in the rearing pond water, characteristic of most intensive rearing situations, supersaturated dissolved nitrogen will not be a problem as long as the total gas doesn't exceed saturation.

Water reconditioning beyond just increased aeration will improve production capacity further. Biofiltration with aeration has produced over 6 kg/l, and such a system has nearly unlimited potential, if sufficient time and space is available to recondition the water. Some farmers presently practice biofiltration by running the water through an aerated shallow ditch partially filled with rocks before returning it to the next set of raceways. Although fish farms may never operate as entirely closed systems, the long-term goal is to greatly increase the efficiency of water use for higher production through oxygenation and biofiltration.

Ammonia removal by aeration hasn't been shown very effective, but that is because at a pH of 6.5 to 7.0 most of the ammonia is in the ionic form (NH_4^+) and not available for gas exchange. Molecular ammonia, however, is subject to the gas exchange laws and will seek equilibrium with atmospheric concentrations. Aeration will remove NH_3, but in low quantities because its form represents such a low proportion of the total ammonia present. However, at a pH between 8.0 and 8.5, which is characteristic of the spring water emerging from the Snake River aquifer, the proportion of NH_3 is much higher, and aeration appears to be more effective in lowering the total ammonia present.

Aeration systems that are designed to maintain dissolved oxygen at 7 ppm throughout the rearing pond may have added value in reducing the ammonia level from the rearing water. If ammonia removal by aeration is to be effective, gas exchange opportunities have to be made available over time because NH_4^+ is in equilibrium with NH_3, and as NH_3 is removed the ionic form will equilibrate by forming more molecular ammonia. If the only opportunity for gas exchange is at the drops between ponds, NH_3 removal won't be very effective. Aeration systems that continually introduce air down the length of the pond will enhance the opportunity for ammonia removal substantially.

Chapter 3

PACIFIC SALMON CULTURE FOR STOCKING

John D. McIntyre

TABLE OF CONTENTS

I. INTRODUCTION

A. GENERAL

This chapter addresses culture of fish to supplement natural production of anadromous Pacific salmonids (*Oncorhynchus* spp.). Many of the techniques involved during the hatchery phases of production are also applicable to the production of fish for stocking net-pens or upland growout production sites.

Waters are stocked with hatchery fish when the native stock does not produce sufficient offspring to use all of the available food and space. Deficits in natural spawning can occur when overfishing or conditions in the migration route, such as inadequate passage facilities at a dam, cause excessive mortality. The primary difference between fish culture for stocking and for other purposes (e.g., ocean ranching) is that in the former instance the fish are to be restored to a native stock either as juveniles or as returning adults from smolts released in the homestream of the supplemented stock.

After some disappointing results, early salmonid culturists were most successful when they released smolts. Because smolts released to waters near a hatchery tend, as adults, to return to that hatchery, there is little interference with native stocks. Some returning fish stray onto natural spawning areas, but they generally are few and the perception in most situations is that interactions with native stocks are insignificant.

Fish production to overcome perceived deficits in natural spawning, however, is a different concept of hatchery operation. The emerging concept is that hatchery fish can be used to supplement natural spawning by increasing the population to a level needed for maximum smolt production.

Natural production can be supplemented with fry, fingerlings, smolts, or even adults. The management goal is to attain and maintain production of a number of smolts that is commensurate with the available food and space. Hatchery fish, however, tend to become a distinct stock because they adapt to conditions in the hatchery. If they are permitted to diverge genetically from the original stock and to interbreed with wild fish, they produce offspring with relatively low fitness for life in local habitats. As might be expected, offspring of hatchery fish do better than offspring of wild fish when both are reared in a hatchery.[221-223] Consequently, it may be impossible to manage these combined systems without some compromise of production from the hatchery, nature, or both. The problem is not unique to fish culture.

> Forest tree management must be something like fish rearing, in a genetic sense. It is impossible to raise a completely natural population under artificial conditions. We have it in our power to cut down every tree on earth, but we cannot replace a mature forest. If we must raise trees, we are forced to sample the genetic variability of natural populations. Our samples are necessarily restricted and cannot duplicate the genetic variability of the species. The same problem must face the hatchery man raising fish. I see no way that the natural diversity could be reproduced under artificial conditions. We can probably do better than we have done, but the replacement of natural populations by restricted samples leaves us vulnerable to pests, diseases, climatic stress, and other hazards.[224]

Although most managers of anadromous Pacific salmonids are not expected to raise a natural population under artificial conditions, they generally are expected to raise a sample of individuals from a population for a time and return them to the population. Management goals are to supplement depressed populations in ways that do not result in unwarranted disruption of their fitness. Some methods are described in this chapter that have been recommended to help reduce the influence of hatchery fish on fitness of supplemented populations. The chapter begins with a discussion of stock structure, stock dynamics, and stocking, and is followed by an overview of techniques for culture of salmonids in hatcheries and the quality of fish needed for stocking.

B. DEFINITIONS

Pacific salmonids are widely distributed species and are composed of stocks or groups of interbreeding individuals. A stock is

> ...the fish spawning in a particular lake or stream (or portion of it) at a particular season, which fish to a substantial degree do not interbreed with any group spawning in a different place, or in the same place at a different season.[225]

and:

> What constitutes 'a substantial degree' is open to discussion and investigation, but I do not mean to exclude all exchange of genetic material between stocks, nor is this necessary in order to maintain distinctive stock characters that increase an individual's expectation of producing progeny in each local habitat.[225]

This definition is widely accepted among fishery biologists and applies to usage in this chapter. Population, as used here, is a group of individuals. Accordingly, a population may be the individuals in a stock or the individuals in a species that has many stocks. Population dynamics is equivalent to stock dynamics.

Fitness is used here as the capacity of a stock to persist over time. This capacity depends not only on the genetic variability in a stock, but also on the genetic system that the stock represents. Genetic systems of locally adapted populations (stocks) often seem to represent an integrated, coadapted array of genes that can be disrupted by stock hybridization.[226] Loss of fitness from hybridization is known as outbreeding depression. Maintenance of a stock's vigor and potential for evolutionary change (fitness), therefore, requires protection not only of its genetic variability, but also of its bases for local adaptations including any coadapted gene complexes.

II. STOCKS, STOCK DYNAMICS, AND STOCKING

Salmonids tend to produce locally adapted stocks.[227,228] Stocks are equivalent to demes[229] in population biology.[230] Traits vary widely among stocks within a species' range[225] and provide a structure that is favorable for evolutionary change. Genetic variability in a stock seems to be the result of heterogeneous micro-habitats and annual changes in physical and biological environmental factors.[230] Small marginal populations evolve that are adapted to relatively extreme environmental conditions,[231] but they may be susceptible to inbreeding and loss of fitness. In spite of these potential problems:

> ...it is in these marginal populations that unique adaptive traits are to be found. They need to be preserved. Marginal populations of fish are a valuable asset for the future.[231]

Species that have persisted have done so despite an occasional loss of a stock, or reproductive failures caused by harsh events such as droughts, floods, earthquakes, volcanic eruptions, glaciation, or disturbances and habitat alterations by man. In a genetically diverse fish population, spawning by strays from a nearby stock could be expected to initiate restoration of a lost stock. Spawning by the phenotypes best suited to surviving a catastrophe can be expected to recolonize the area.

Subtle adaptations for life in waterways characterized by specific temperature, chemistry, altitude, and species assemblages are difficult to identify, but hypotheses concerning the presence of such adaptations generally are supported as the data become available.[232-235] Adaptation involves an intricate shuffling of interacting genes into a network of effects (a "coadapted gene complex").[236-238] Different environments induce the development of different gene network configurations. Populations adapting to similar environments, but independently and in isolation from each other, may develop different complexes.

The mingling of two different gene networks (mixing stocks) may disrupt the effectiveness of either, even if the two gene networks independently adapt their respective populations to similar environmental conditions. From data for 22 populations of organisms used in experimental crosses, viability in 19 of those crosses was reduced in the second generation and indicated disruption of coadapted gene complexes.[237]

Genetic diversity affects survival and production because requirements for survival in a particular environment can be stringent. We seldom have enough information to identify such effects. However, one example has recently been described. Wild coho salmon, *O. kisutch,* that persist in Oregon's Nehalem River have genotypes that enable them to survive in an environment with a virulent parasite, *Ceratomyxa shasta.*[239] Fish in the North Fork of the Nehalem River remain sensitive to the parasite, presumably because they have not had to tolerate its effects. The gene pool in each stream has provided the traits for survival and growth in the respective environments. Genetic change during even a few generations in a hatchery seemingly suffices to alter survivability. Survival of steelhead, *O. mykiss,* from Oregon's Round Butte Hatchery stock was lower than that of native fish when both groups were reared in tributaries of the Deschutes River,[221] even though the hatchery stock was developed from the wild stock. Growth and survival were greater for hatchery fish than for wild fish when offspring from both groups were reared in Round Butte Hatchery.

When large differences between hatchery and wild fish exist, the observed effect on survival is more pronounced.[221] In a study of the reproductive success of hatchery and wild steelhead in the Kalama River, Washington the hatchery fish originated from a stock other than the native Kalama River stock.[240] The hatchery fish were only 28% as successful in producing smolt offspring as were the native fish. Although the authors could not separate the effects of adaptation to the hatchery from those of differences in origin, they did conclude that interbreeding of the hatchery and wild fish put the fitness of the latter at risk.

Similarly, production of juvenile coho salmon in streams stocked with hatchery fish was lower (average reduction of 32%) than in unstocked streams.[241] Stocked streams produced no more adult fish than unstocked streams. The authors concluded that fitness of the hatchery fish was low because they spawned at an inappropriate time.

Even without selective mortality in a hatchery there is still an insidious effect of fish culture[242] because hatchery conditions generally lead to high survival in an average brood. Many young fish that would have been removed from the population by natural selection in a stream or lake are protected in a hatchery. Once released, those fish may survive conditions in the ocean and return to spawn. If their offspring are also protected from natural selection pressures in a hatchery, the frequency of fish that are maladapted for life in a relatively perilous natural environment will increase in the population.

Experimental results have confirmed expectations based in theory and showed relatively poor survival of the offspring of hatchery fish in natural habitats. Hatchery fish that would have been removed by natural selection may be added to a native population displacing native fish with fish that are not adapted for survival in the natural environment. Further, if they survive to reproduce, they may produce offspring that are not fit for survival in nature. Because of the risk to the fitness of native stocks resulting from stocking, hatchery programs that are part of a program to ensure the persistence of salmonid gene resources must be carefully developed and managed. The fish produced in hatcheries must not have characteristics that will exacerbate the tendency for stocking programs to disrupt fitness of the native fishes.

The number of individuals in a population of juvenile anadromous salmonids in freshwater depends on available food and space.[83,243] Densities that would result in deleterious "stunting" of individuals in stream-dwelling populations, and perhaps lake-dwelling populations as well, seem to be prevented by emigration and mortality. Individuals that cannot successfully defend a territory presumably emigrate in search of less hostile conditions. Emigration and mortality increase as density increases so their effects are density dependent. Irrespective of population size, variable climatic conditions result in greater productivity of streams for salmonids in some years than in others. These differences also affect growth and survival, and population density in any one year is determined by the interplay of density-dependent and density-independent effects.[244]

Stocking with hatchery fish tends to increase density-dependent mortality in fish produced by natural spawning. In addition, at least some of the wild fish will also be displaced if hatchery fish are given an advantage in behavioral interactions because of large body size or other traits. Presumably, losses of wild fish from increased mortality and displacement by hatchery fish would be insignificant if the fish were genetically similar. The degree of similarity depends in part on the source of brood fish, the character of the hatchery operation, and the relative number of hatchery and wild fish in the population.

Density-dependent mortality during the egg stage can result when large numbers of fish spawn in a limited amount of space and superimpose their redds upon one another. In later stages of development, intraspecific competition for limited food and space can result in a declining rate of increase in smolts produced as spawner density increases. At high spawner or juvenile densities, smolt production may even decline.

Smolt abundance in populations that are regulated largely by density-dependent processes can be predicted from a relation describing the number of offspring produced by increasing numbers of adults. Relations between parents and smolts may not be obvious for some populations, however, indicating that density-independent factors are of primary importance in determining smolt yields for these populations. Water temperature, for example, may vary to the extent that survival of juveniles is effectively controlled by heat or cold, and annual yield of smolts varies independently of parent abundance. If density-independent factors are of primary importance in limiting population size, presmolt stocking would be successful only in exceptional situations where the limiting mortality occurs early in the life history and subsequent conditions are conducive to survival.

Because mortality increases with increasing fish density, stocking with fry or fingerlings can cause an increase in mortality in the supplemented population. The ratio of hatchery to wild fish in the resulting smolt population will reflect the proportions of each in the supplemented fry population if the hatchery fish had no survival advantage or disadvantage. If, on the other hand, the hatchery fish have an advantage, such as being larger, the ratio of hatchery to wild fish in the smolt population will favor the hatchery fish. In either case, wild fish will be displaced (through increased emigration or mortality). A remnant stock can be extinguished if it is continually faced with high mortality because its members must compete in a population that is inflated in size with fish from a remote stock.[245]

It is conceptually possible to manage a stocking program to attain balance among the deficit in natural spawning, the number of hatchery fish to plant, and harvest mortality.[246] More likely, however, stocking will continue on the basis of fish availability, manpower, equipment, and accessibility of planting sites. Increased density-dependent mortality from stocking will probably be disproportionately greater for wild fish than for hatchery fish because of a size advantage in the hatchery fish. Consequently, protection of genetic variation in anadromous salmonid resources increasingly depends on production of fish for stocking that do not disrupt the gene pool of a supplemented stock.

III. FISH FOR STOCKING

A. CULTURE METHODS

Pacific salmon and anadromous trout generally return for spawning to the locations in freshwater where they lived immediately before emigrating to the ocean. This behavior facilitates convenient capture of adults. The adults do not return or mature en masse, so hatchery workers must be prepared to capture and spawn brood fish several times during a given fish run.

Most species or races of anadromous salmonids entering freshwater rivers and streams in the spring or summer (e.g., spring and summer chinook salmon, and summer steelhead, but excluding the winter chinook salmon, *O. tshawytscha,* in California's Sacramento River[54]) are not sexually mature. For example, spring chinook salmon may return to a hatchery in June, but may not mature sexually until September. Summer steelhead may return in June, and not be ready to spawn until the following spring. If captured soon after they enter freshwater, adults must be maintained in ponds with sufficiently good water quality and cover to ensure the maturation process continues.

Some species and races enter freshwater in the fall or winter (e.g., fall chinook salmon, coho salmon, and winter steelhead) and are sexually mature when they arrive at the spawning grounds. Those collected in traps are generally near maturation and can be spawned soon after capture. Differences in timing among taxonomic units persist, but are less pronounced in the northern portions of a species' range.

Because not all are ready for spawning at the same time, adults must be sorted on a regular basis. Brood stock holding ponds are constructed so fish in a pond can be crowded into a relatively small area for capture with a dip-net. Once a fish is in hand, gentle pressure applied to its abdomen will result in the expression of fully developed eggs or milt (seminal fluid and semen). Immature fish are returned to the holding pond. The procedure is repeated at intervals of several days until the desired number of fertilized eggs is obtained or the supply of brood fish is exhausted.

Salmon are killed before spawning by a blow to the head. Some culturists sever a female's caudal peduncle to drain much of the blood volume away before it contaminates the gametes as they are removed. Eggs are permitted to fall into a pail through an incision made in the female's abdominal cavity. Milt is expressed into the pail by applying pressure to the abdomen of a male.

In some hatcheries, steelhead and cutthroat trout (*O. clarki clarki*) brood fish are not killed for spawning. Unlike salmon, they are iteroparous and may survive to spawn again. Adults of those species may be anesthetized, and their eggs or milt expressed by gentle pressure to the abdomen. Spawned fish are returned to freshwater for recovery and later released to natural waters.

The combined eggs and sperm are gently stirred, and water is added to activate the sperm, and presumably facilitate fertilization. After the fertilized eggs are rinsed in clean water to remove excess fluids and tissue they are put into an incubator. Many types of incubators are used, with fiberglass trays arranged in stacks being one of the most common.

Incubating embryos are permitted to develop undisturbed to the eyed stage. They are then shocked to rupture the yolk membrane, causing the yolk to coagulate. White eggs are dead and easy to identify and remove. An adequate shock can be produced by siphoning the water and eggs from a trough and letting the eggs free fall 50 to 100 cm into a pail. Dead eggs must be removed at regular intervals until the embryos hatch because the fungus *Saprolegnia* sp. invades the dead tissue and, if permitted to expand, may attack live embryos.

Egg picking machines have been developed that will automatically separate viable eggs from those which are dead. Eggs are passed single-file through a light beam. If the light passes through an individual egg, that egg is retained. Dead eggs are opaque and when they pass in front of the light beam they are expelled from the machine into a collection bucket with a puff of air. Egg picking machines are efficient and can save large numbers of employee hours.

The rate of larval development depends on temperature. Culturists have developed relations between temperature and the elapsed time to specific stages of development. Heat is quantified in three ways:

1. As daily temperature units (TU or DTU), or the degrees (in Fahrenheit) that exceed the freezing temperature in a 24-hour period;
2. As monthly temperature units (MTU), or the average daily Fahrenheit temperature for a month minus 32; and
3. As degree days.

The TU's required to attain specific stages of development are species specific and seem to be slightly affected by fluctuating temperature, but the same approximate relation persists as with constant temperature.[47] Available data are insufficient for accurate descriptions of the relationship which holds at low temperatures.

After hatching, alevins are retained in the incubator until they deplete their yolks. Free-swimming fry have to be fed. They are removed from the incubator, put into troughs or tanks, and fed a starter diet. At a length of 2.5 to 3 cm, fry are moved to larger troughs or raceways to complete development in the hatchery.

Dates and sizes at which fish are either released from the hatchery or hauled to a remote release site, depend on the management goal, water temperature, hatchery space requirements, behavior and condition of the fish, experimental results, and flow conditions in the receiving waters.

Development of hatchery procedures has eliminated many problems that plagued early culturists. Nevertheless, culturists must continually guard against sporadic outbreaks of disease or rapid changes in water quality.

Readers requiring more information are referred to the chapter on Atlantic salmon in this volume (Chapter 5), Piper et al.,[45] Leitritz and Lewis,[44] and Bardach et al.,[247] along with such journals as *The Progressive Fish-Culturist* and *Aquaculture*. Senn et al.[248] summarized a great deal of useful information and reference material for hatchery managers, hatchery construction engineers, and fish culturists.

B. FISH QUALITY

Salmonid hatcheries and culture techniques have been subjects of concern and study for several decades because of their effects on the produced fish. Early conclusions[249] were that survival of hatchery fish was low because of excessive fat and carbohydrate in their diet, overfeeding, lack of sufficient exercise, inability to forage or avoid predators, acclimation to stable water temperatures, increasing domestication, artificial selection for some traits, absence of natural live food in their diet, and poor transportation and planting techniques. Additional effects on morphology, development of respiratory systems, maturation rates, the smoltification process, feeding efficiency, imprinting to rearing sites, and behavior have been described more recently.[232] The authors pointed out that the hatchery environment has important effects on survival after release, and also affects the gene pool.

As stated, the goal of fish culture for stocking is to produce fish that will not disrupt the fitness of the supplemented stock and will do what is expected of them after they are released. Released fry or fingerlings are expected to substitute for fish in the native stock and, soon after release, become an integral part of the locally adapted population. Hatchery smolts should have at least the same potential for survival as their wild counterparts and are expected to return as fully capable members of the natural spawning run at the release site. Accordingly, both genetic and environmental considerations exist for production of good quality fish for stocking. Some methods and concepts to help culturists manage both the genetic and environmental components of fish quality to meet stocking goals follow.

The source of brood fish is important to the success of a hatchery program. Fish from distant stocks do not survive as well as fish from local stock; survival tends to decline with the transfer distance.[223,250] Returns of transferred fish may be 5- to 10-fold lower than returns of local fish.[223] Many attempts have been made to improve native fish in an area by transferring fish with the desired characteristics. Such practices may seem logical in the absence of further information and may in fact work as expected. However, there is no reason to expect the desirable characteristics to persist once the stock is transferred, or even to expect that the stock itself will persist at the new location. Failure should be expected because any one gene influences many traits, and any given trait is almost invariably controlled by several genes.

The argument that genetic diversity will be enhanced by stock transfer from distant populations is valid, but the accrued advantages will probably be strongly reversed because of resultant destruction of locally attuned, coadapted gene complexes. Sufficient diversity can be maintained by avoiding extensive inbreeding and can be accomplished in wild and natural stocks by allowing only a low frequency of natural strays. In hatcheries, the problem can be avoided by using sensible breeding practices.

Accordingly, the initial brood stock at a new station should be taken from the stock that is to be supplemented. The risk of disrupting fitness in the native stock is low if brood fish in subsequent generations are also from the natural spawning run. The risk is greater, however, if hatchery fish beget hatchery fish because they will tend to diverge genetically from the native stock. Development of a hatchery stock that is isolated from the native stock brings potential problems (inbreeding and genetic drift) with small population size into the hatchery management strategy.[202,251-253] Members of a population with closely related ancestors share genetic material and may experience inbreeding depression even if their population is large. The spawners in a small population tend to pass a biased sample of the gene pool to the following generation because small samples often do not represent the population from which they are taken and cause change (drift) in the gene pool.

Inbreeding and genetic drift may not be a concern for most managers of hatcheries that produce fish for supplementation because those managers are not conducting a captive breeding program wherein hatchery fish beget hatchery fish without interbreeding with fish from native stocks. Because anadromous fish generally produce strays between adjacent stocks and at least a few wild fish are included in each brood of hatchery fish, inbreeding and genetic drift effects would not be expected. Genes introduced by even a single migrant per generation may swamp the effects of inbreeding,[254] regardless of the effective population size.[202,253] Problems may arise, however, in the presumably rare situation when a stock is isolated from interbreeding with strays from other stocks. A minimum effective population size of 500 is needed to protect the historical level of genetic variation in an isolated native population.[255]

Once steps are taken to ensure that local adaptations are maintained and problems of small population size are addressed, loss or change in the genome of the hatchery fish must be prevented. First, managing a brood of hatchery fish without influencing its gene pool is probably impossible. Suppose, for example, that photoperiod control was used to stimulate brood fish to spawn earlier in the year than they would under a natural photoperiod. This is a change caused by environment, but it puts the offspring of the fish involved in conditions other than those to which their ancestors had become adapted. Natural selection will favor individuals given some advantage by conditions that exist during the new spawning period, rather than individuals adapted to conditions in the normal spawning period. Change seems inevitable, but managers can take steps to ensure that hatchery fish do not cause unwarranted disruption in fitness of supplemented native stocks.

All traits of a stock are the result of the evolutionary process. Some traits are fixed (presence of eyes, fins, etc.) but others change as environment changes in space and time (e.g., run timing, body size, age at maturity). Traits that can easily be changed presumably enable species to persist in variable environments. If conditions change so that selection pressure always favors certain phenotypes (e.g., early spawners or large body size) other phenotypes will tend to disappear from the population.

Intentional or unintentional selection can occur at any stage in the life cycle of a population. Retention in a hatchery population of only spawners that possess a desired trait such as large body size is an example of intended selection. Unintended selection can result from any source of mortality. Suppose, for example, that the oxygen in the water supply of a hatchery is depleted for a brief period causing high mortality of fry. Frequencies of genes associated with low tolerance to anoxia will be low in the survivors of such a catastrophe compared to frequencies in the population before the die-off. The chance that specific genes or gene complexes will be lost, however, depends on the proportion of the population killed in a catastrophe and the frequency with which the catastrophe occurs in subsequent generations. A catastrophe that kills 20% of a population in one generation and is repeated once in every ten generations, for example, is less likely to result in loss of genetic material than is a catastrophe that leads to 80% mortality in a population each generation.

Disease problems are an ever-present threat to the success of hatcheries. As in any situation with artificial high densities of animals, sporadic epizootics will occur. These events represent not only a significant source of selective mortality for the fish but have caused technicians[256] to resort to methods that result in selection for reproduction of only those fish with immune properties deemed desirable by the technicians.

Effects of selection cannot accumulate in anadromous salmonids produced in hatcheries where the parents for each brood are captured from a large run of wild fish. Effects may accumulate, however, in a run that is mostly hatchery fish and hatchery fish beget hatchery fish. Because, as discussed earlier, selection is counterproductive to the production of fish for stocking, and stocking will replace fish in endemic stocks with hatchery fish, there seems to be no justification for intentional selective breeding in stocking programs. The prudent policy is to do all that is possible to minimize the effects of unintentional selection and to prevent intentional selection unless it serves to overcome nonrandom effects caused by man. Hatcheries tend to produce fish that return earlier in a given year than their wild counterparts. Reversal of that tendency may be a candidate for selection. Selection to overcome the selective effects of some fisheries may be another candidate.

Hatchery managers either have a captive brood stock or yearly access to spawners from a natural population. The former have the greatest and the latter the least chance of losing or altering genetic material in a population. The care, ingenuity, and attention required by hatchery managers to prevent genetic losses is obviously highest in the former. Where hatchery fish beget hatchery fish without gene flow from a native stock, managers can do little to prevent the population from adapting to an environment defined in part by early hatchery experience. Genetic material in the original sample of fish used to initiate the hatchery population will probably be altered by selection in spite of careful management. All efforts to prevent selection are thwarted, for example, when water quality control fails or an epizootic causes high mortality and retention of only individuals that tolerated the stressor. Ensuring 100% survival is not possible and some selection will occur under the best conditions. Unless stress recurs each year, the losses will probably be restored in species with overlapping generations.

In the presumably rare situation where inbreeding is a potential problem, managers must keep the rate of inbreeding as low as possible. The ideal condition for avoiding inbreeding[202] is to ensure that two offspring from each pair of parents are used as parents in the subsequent brood. When one pair produces less or more than two reproducing offspring, their genes are not represented in proportion to their abundance in subsequent broods. Separate families are not identifiable unless marked, but steps are available for preventing unwarranted inbreeding and genetic drift and include use of all fish returning to a station in the brood, mating one male to one female, and, to the extent possible, using each male only once. One might expect to reduce the variance in male family size by fertilizing the eggs from each female with a mixture of the sperm from several males, but the ensuing interaction among sperm used in that manner is to the apparent disadvantage of some males.[257,258]

Sperm may compete for eggs and make the use of mixtures of sperm inadvisable, but use of a single infertile male must result in wastage of available eggs. To overcome such a problem, males can be used in overlapping pairs to fertilize the available eggs.[257] Sperm from males A and B is mixed and used immediately to fertilize the ova from female A. Sperm from males B and C fertilize ova from female B and so on until all adults are spawned. Ova from the last female are fertilized with a sperm mixture of the first and last male.

As described earlier, stocking with fry and fingerlings is direct interference in the dynamics of a stock. The mortality from the increased density is exacerbated if the hatchery fish are able to dominate the wild fish, for example, by larger size. Consequently, the quality of salmonid presmolts for supplementing native stocks without disrupting their fitness depends, in part, on body size. All else being equal, the least harm to native stocks that are to be supplemented with presmolts presumably will occur when hatchery fish are no larger than fish in the supplemented stock. Modern diets and water temperature controls are such that hatchery personnel can produce fry or fingerlings of desirable size.

Although presmolts are used to supplement some native stocks, the use of smolts is becoming more common. Smolts are expected to remain in the stream only briefly before they emigrate to the ocean. Their size relates to quality only in terms of the size required for greatest survival. The size of smolts and the time of year that they are released are two of the primary determinants of their success.[89,259,260] Large smolts tend to survive at higher rates than small smolts, but the relation between size of smolts and return of adults may be species specific and differ by location. The importance of time to smolt migration was made clear at Carnation Creek where a slight change in temperature influenced the time that fish left as smolts, which, in turn, had a profound influence on the number of produced adults.[261]

Some other factors that influence survival of hatchery smolts also seem to be species-specific. Survival of chinook salmon smolts declined as densities in hatchery raceways increased above low to intermediate levels.[262] Returns of chinook salmon to Carson National Fish Hatchery on the Columbia River did not increase at rearing densities greater than 20,000 fish per raceway,[262] but preliminary results with coho salmon do not support a similar conclusion.[263]

Hatchery fish infected with *Renibacterium salmoninarum*, the causative agent of bacterial kidney disease, may die after they are released.[264] The incidence of bacterial kidney disease, however, did not correlate with return rates obtained for chinook salmon at Carson National Fish Hatchery.[263]

Literature on fish culture includes many descriptions of tests to assess hypotheses about the potential role of various behavioral, morphological, and physiological traits as indicators of the quality of fish produced in hatcheries. It seems, however, that the only traits that have been shown to influence smolt survival are the source of their parents, the time of release and size at release, and, at least for chinook salmon, the population density during rearing. Although the relation between incidence of disease and survival after release is not clear, release of fish with debilitating diseases is unacceptable, especially in stocking programs. Possibilities for improvement of survival in smolts by increasing responsiveness to predators[217,265-267] and by improving their physical stamina[268-270] have not been sufficiently explored.

IV. RELATED TOPICS

A. MANAGEMENT UNITS

A first step in protecting the fitness of anadromous fish is to identify, among the fish of a basin, the aggregations to be considered as separate units (stocks) for conservation management. It is no longer realistic to presume that the structure from natural evolutionary processes can be maintained.

Outbreeding depression may not occur between closely related groups.[202,271] If those findings are correct, development of management units that include formerly closely related stocks is possible without causing unacceptable losses of fitness. The geographic distance between groups of anadromous salmonids that are to be combined cannot be great, however, because fish from adjacent tributaries of the same river can have site-specific adaptations.[239] Such findings should not deter development of conservative management strategies, however, because an exhaustive survey of specific adaptations takes effort and much time, and the result may not be conclusive.

Definitions of stocks based on, for example, biochemical, morphological, or physiological types in populations have been attempted, but none has provided a basis with sufficient detail for separating individual breeding units and for describing the relationships among groups. Gene frequencies determined from biochemical variants, for example, do not permit separation of some groups that have significant differences in habitat preferences and life histories (e.g., summer and winter races of steelhead).[272] Given the limitations of technology, an alternative system for classification must be devised that uses workable, objectively defined criteria. Upon completion of a classification system, protection of fish in each unit from unwarranted interference from fish in other units may be developed. Brood stock for a unit can be obtained from adults reared from young in that unit.

The basin or subbasin with an anadromous species provides a useful basis for classification. Populations can be defined on the basis of stream order. The smallest permanent streams are order 1. Order 2 streams are formed by the confluence of two Order 1 streams and so forth, up to maximum orders of 7 or 8 for the mainstream of most large rivers. Managers could use this system to decide at what level (stream order) to protect local adaptations. At a 1:500,000 scale, for example, about 36 order 5 and 146 order 3 streams exist in the Columbia River basin according to data available from the Northwest Power Planning Council. Populations so defined probably include several smaller stocks, and combining small stocks into a single larger population for management as a single genetic unit may reduce fitness for the species in the basin. That is, some loss of fitness may be an inevitable result of this crude classification scheme. In the absence of such management units, however, the existing practice of regularly transferring fish from area to area within a basin may cause much greater disruptions in fitness.

Given the uncertainty about the basis for local adaptations and the geographic area over which they are important, a conservative program is in order. Management schemes to protect supplemented native populations should protect as many stocks or aggregations that will be treated as stocks in as many geographic areas as possible. Neither hatchery nor native stocks should be transferred across the established boundaries for a management unit. After several generations of fish have been treated as a separate group under such a classification system, their genetic structure will probably be consistent with Ricker's definition for stock.[225] Managing to protect differences among those stocks may be even more important to maintaining fitness of a basin's population than managing hatchery fish to prevent loss of genetic variants.

B. FISH QUALITY

The use of hatchery smolts to supplement a native stock presumably does not pose a direct threat to the native stock at the time of planting because smolts begin their seaward migration soon after they are released. Smolts are intended to produce adults that will spawn near the planting site and, therefore, must be genetically compatible with fish in the native stock. Their size and behavior at release are not expected to disrupt fitness in the native stock.

Stocking with fry or fingerlings (presmolts), however, must be with fish of appropriate genetic make-up and no larger than fish in the supplemented population. If they are larger, they can be expected to displace a disproportionate number of native fish, thereby increasing the risk to fitness of the native stock.

C. COSTS OF FITNESS PROTECTION

Because of genetic make-up and environmental history, some phenotypes are expected to thrive and some are expected to perish in a hatchery. In a different environment, even another hatchery, an entirely different group of phenotypes may thrive. Genotypes favored by this selection persist in a hatchery. The result of these selective forces is a population of individuals that, compared with the parental population, have increased fitness for life in a hatchery and reduced fitness for life in natural habitats.

Conservation managers must acknowledge that the best fish for stocking may not be individuals with the greatest potential for survival after release because they have not been permitted to adapt to the hatchery. Likewise, there is some cost to a hatchery, in terms of fish survival, of managing supplementation so that fitness of native stocks can be protected.

D. STOCK TRANSFER

Transfer of stocks is costly. It may result in a serious risk to the fitness of native stocks (and to hatchery stocks). Some authors[273] argue that poor performance of stocks from genetic deterioration can be overcome by hybridization, but it was pointed out for wild stocks that such an approach is "...genetically and ecologically naive and the results potentially dangerous."[274] Economically, it is senseless to use fish only because they are easily obtained from some remote rearing station with an excess supply on hand. Until information is developed to show the range over which specific brood stocks can be used (transported) without disrupting fitness of locally adapted stocks, managers have to prevent transfer and interbreeding of fish from different locations.

E. CONSERVATION VS. ENHANCEMENT

Some hatchery stocks (enhancement hatcheries) are maintained only to produce smolts that, upon release, are expected to emigrate directly to the sea. As adults, those fish are expected to be captured or to return directly to the hatchery where they were spawned. This is the strategy for coho salmon and fall chinook salmon in hatcheries downstream from Bonneville Dam on the Columbia River. Because conservation of natural gene pools is not an issue (endemic stocks have no doubt been lost as a result of the combined effects of overexploitation and straying by hatchery fish), these hatcheries might be considered candidates for the use of genetically engineered fish as the technology becomes available.

In spite of best intentions, managers cannot ensure that no differences will be produced between fish in the hatchery and fish in natural parts of the system. For example, many fish that good culture techniques preserve in the hatchery would die in nature, and measures to prevent epizootics will no doubt continue to exert novel selection pressures in hatcheries. Although survival of hatchery fish and their offspring upon release from the hatchery is less than for fish produced in nature,[222,275] intensive management[276] to prevent systematic genetic changes in these fish probably can protect genetic variation needed to ensure persistence of stocks.

F. SUMMARY

Some brood selection strategies and culture methods that are helpful in prevention of disruption of fitness in native stocks are addressed. Similar summaries exist in several of the references previously cited in this chapter.

1. Managers can ensure that culture practices do not change adaptive traits such as the time that adult fish return for spawning.
2. Growth rates can be manipulated via temperature and diet control so that planted fish are not significantly larger than fry or fingerlings in the natural population.
3. Adults of all ages and from all parts of the maturation cycle can be included in the brood stock each year at the proportion in which they exist in the run (unless selection is needed to protect the timing of the run).
4. Steps can be taken to ensure that slow growing fish are not discarded so that rapid growth does not receive a premium.
5. Individual males can be prevented from having an advantage in reproduction as a result of hatchery practices.
6. Selection based on appearance can be prevented.
7. Selection can be initiated to overcome the selective effects of some fisheries.
8. The initial brood fish at a new station should be taken from the run to be supplemented and subsequent generations should consist of adults from natural spawning in numbers reflecting their relative abundance in the run.
9. Stocking of eggs, fry, or presmolts should proceed with knowledge of the deficit in the population from natural spawning to prevent overstocking and cause excessive density-dependent mortality for juveniles from natural spawning.
10. Stocking with fish from a hatchery should be limited to the geographic area inhabited by the stock from which the brood fish were obtained.

ACKNOWLEDGMENT

I thank Barbara Hayden for editorial assistance and John Emlen for many hours of helpful discussion. Mary McIntyre and Reg Reisenbichler helped in many special ways. Al Fox thought that my concept of stocking and fish for stocking should be described.

Chapter 4

GROWOUT OF PACIFIC SALMON IN NET-PENS

Robert R. Stickney

TABLE OF CONTENTS

I. INTRODUCTION

Development of the technology under which salmon could be reared in floating net-pens began in the late 1960s in Washington state.[277] A 1970 report,[278] which may relate to the first salmon net-pen constructed in Puget Sound, describes a structure of approximately 58 × 14 m which contained a net that was about 3 m in depth. The pen was constructed by the Bureau of Commercial Fisheries — which is now the National Marine Fisheries Service (NMFS) of the U.S. Department of Commerce — and located near Anacortes, Washington. Research in net-pens with volumes of 2.7 m^3 began at the Manchester laboratory of NMFS in 1969.[277,279] Those early studies led to the development of a larger research facility at Manchester[279] which continues to be in use (Figure 17).

Cage culture for trout production is a variation on the net-pen theme which has been evaluated for its potential in freshwater impoundments.[280] Cages used for the culture of fish in freshwater are generally smaller than marine net-pens. The cage concept actually preceded that of net-pens by perhaps as much as nearly two decades and has been applied more widely to warmwater fishes such as catfish than to salmonids.[281] A typical freshwater fish cage tends to be in the range of 1 m^3, though those used by Boydstun and Hopelain for rainbow trout[280] were 12.2 × 2.9 m in area and 1.5 m deep (53 m^3 in volume).

Initial work in Washington net-pens focused on the rearing of coho and chinook salmon. One of the aims of the program was to select rapidly growing fish for use as brood stock.[279] The first coho salmon were reared to maturity in net-pens by NMFS researchers by the early 1970s.[282-284] Chinook salmon were first reared to maturity in net-pens at 3 years of age, but they were infertile.[277]

An advantage of chinook was that they could be transferred to the marine environment early in their life history, but a problem with scale loss during the first year led to a high incidence of infections.[279] The author expressed the view that the diets which were being fed were inadequate and contributed to some of the observed problems. Among the diseases which appeared during the first few years of culture in net-pens were vibriosis, furunculosis, and bacterial kidney disease.[277]

The first commercial net-pen facility in the United States was the Domsea farm which was located adjacent to the Manchester NMFS site. It was developed in 1971[279] after the state of Washington approved a statute which removed a prohibition against private salmon farming on a limited basis.[285] Under the legislation, the farmers were allowed to purchase eggs from the state for a period of six years, during which they would be expected to develop their own brood stocks. The 6-year period was extended indefinitely when it was learned that brood stock development required a longer time frame. Currently, egg sales to Japan from the states of Washington and Oregon exceed 50 million coho and chinook, while Washington fish farmers purchase few eggs, having redirected their emphasis primarily to Atlantic salmon production.[4]

The Domsea facility ultimately covered some 2 ha with net-pens (Figure 18), and holds the record as the largest net-pen facility in the world. The pens, numbering 250, were 7.3 × 14.6 m and 3.6 to 4.4 m deep. Coho salmon were reared to pan-size in the pens. A total of 10 to 15 million eggs were required annually to produce fish for stocking the facility.[285] Reconstruction of the facility was initiated in 1988. That activity involved downsizing the number of net-pens and replacing them with more modern structures. The operation, now under new management, is converting from coho and chinook to Atlantic salmon production.[4]

Permitting constraints slowed the development of salmon farming in Washington state. As of early in 1990, there were still only 13 commercial salmon farms in the state,[141] though permits for a few additional farms were under consideration. Major development of salmon

FIGURE 17. The net-pen facility of the National Marine Fisheries Service laboratory at Manchester, Washington is used for the conduct of research. (Photo by Robert R. Stickney.)

net-pen culture between the early 1970s and the late 1980s occurred not in the United States, but in Norway, Scotland, Japan, and a few other nations, with the primary species of interest in many countries being Atlantic salmon. In the past few years, Pacific salmon net-pen culture, along with that for Atlantic salmon, has been initiated in Chile and in British Columbia, Canada. Atlantic salmon is the species of choice in eastern Canada. Culture of Atlantic salmon is considered in Chapter 5 of this book.

The salmon farming industry in British Columbia began in 1979 and grew fairly rapidly until 1982 when it peaked at slightly over 270 tons.[286] In 1983 and 1984, production fell to 100 to 125 tons. A similar rise from 1976 to 1980 occurred in France, where the primary species under culture was rainbow trout, along with a modest amount of coho salmon and brown trout. A production plateau was reached in France at near 450 tons, where it remained through 1983.[287]

A moratorium on new salmon net-pen farms was imposed in British Columbia for a period, but was lifted in 1986 when the provincial government decided to actively promote aquaculture development. On March 20, 1986, there were 40 active salmon farms in British Columbia.[286] In addition, 69 new applications had been approved and 108 were pending. By early September of the same year, there were 85 active farms and over 700 applications were pending. Not all of the applicants actually went into production. Incredible growth

FIGURE 18. The Domsea net-pen facility, located near the National Marine Fisheries Service net-pens at Manchester, Washington, was the largest such complex in the world. (Photo courtesy of Ronald W. Hardy.)

occurred between 1986 and 1988.[288] In 1988, there were approximately 125 farms which actually had fish in the water.[286]

The incredible increase in net-pen salmon farming activity in British Columbia, in association with the development of Atlantic salmon farming in eastern Canada, and the continued growth of the Atlantic salmon farming industry in Europe, culminated in a major restructuring of the industry in 1989. Sagging prices, particularly with respect to chinook salmon,[288] excessive debt levels, and high interest rates, in conjunction with a saturated worldwide market, led to significant impacts on British Columbia salmon producers.[289] While the 1989 production forecast was for 14,500 tons, the price collapse led to the marketing of 12,400 tons at prices 30 to 40% lower than projected.[288] The farm gate price of fresh farm-raised salmon dropped to $6.60 Canadian per kg, whereas production costs ranged from $5.50 to $7.70 per kg. By May 1989, two farms had been sold at bargain rates, at least four were in receivership, and many others were in serious financial difficulty.[289] As the industry recovers from the economic crisis of 1989, it might be expected to grow at a slower pace in response to market demand and because lending institutions may be less willing to provide venture capital.

Chinook salmon is projected to remain the dominant species in British Columbia net-pen culture.[288] In 1990, chinook comprised 77% of total production in the province. Coho salmon production has been declining and was projected to comprise only 7% of production in 1990. Atlantic salmon make up the remainder. The production estimate for British Columbia during 1990 has been placed at 15,000 tons.[288]

While salmon farmers in British Columbia and the United States are turning increasingly toward the Atlantic salmon, the coho salmon farming industry in Japan continues to grow. It was projected that farmed coho salmon production in Japan would top 20,000 tons in 1990.[290]

II. CULTURE PRACTICES

A. NET-PEN CULTURE SYSTEMS

The design of typical net-pen facilities employed in Puget Sound, Washington has recently been described.[141] The basic component is a group of pens that are supported in a manner which allows them to float at the surface of the water. While the actual number and size of net-pens associated with a particular farm varies (the current limitation in Washington requires that net-pens cover no more than 8,100 m^2 at a permitted site), a typical farm will consist of up to 50 pens laid out in a square or rectangle. A traditional farm might be 30 × 300 m in size. Pens are usually rectangular with common sizes ranging from 3 to 6 m wide and 6 to 12 m long.[291] Circular pens up to 30 m in diameter have also been used in clusters of three to four per facility.[141]

Supporting structures for net-pens have been constructed from galvanized metal, plastic pipe, and wood. Many commercially available net-pens have frames constructed from galvanized steel (Figure 19). Walkways of 1 to 2 m width provide access to all sides of each net-pen. The nets are typically hung from 1-m high railings that are located along the inside edges of the walkway. By hanging nets from the railings it is possible to keep them well above the water line to prevent fish from jumping out of the enclosure. Nets generally have 1- to 3-cm mesh, depending upon the size of fish being held.[141] A typical net will be from 5 to 8 m deep and provide a total volume of 40,000 to 65,000 m^3. Bird netting is often suspended over the top of each net-pen to deter predation.[291]

The net-pen structures may be supported by flotation devices such as styrofoam blocks, but most modern facilities feature a system of external floats (Figure 20). A typical mooring

FIGURE 19. Commercially available net-pens feature galvanized steel walkways. The nets have yet to be hung on the structure shown here. Note the floats distributed under the walkways. (Photo by Robert R. Stickney.)

system involves attaching floats to the structure at various points. A long cable, chain, or rope is then attached from each float to an anchor weighing from 1,360 to 2,270 kg.[141] The floats hold the net-pens in position relative to the waterline and help moderate the effects of inclement weather and the currents associated with winds and tides.

Some net-pen facilities have walkways which provide direct shore access, but many are sited sufficiently away from the shoreline that they must be serviced by boats. In some instances small buildings have been constructed on the net-pen structures to store feed and supplies, but it is common practice to keep support facilities ashore and ferry feed and other materials to the net-pens as required.

B. SITING CONSIDERATIONS

Fish produced in net-pens require water of high quality (see Section D below). The presence of the fish themselves will lead to increased ammonia levels and reduced dissolved oxygen, at least in the immediate vicinity of the net-pen, unless sufficient currents are present to provide rapid water exchange. In a well-flushed net-pen, the only change in water quality that can be detected should be within the structure itself.[292]

Currents will be slowed within net-pens because of the drag provided by the netting and support structures associated with the facilities. In order for proper flushing to occur, a current velocity of at least 10 cm/s should be available outside of the net-pen structure throughout most of the tidal cycle.[293-295] Current velocity can also be too high, leading to stresses on the culture system and requiring the fish to continuously swim rapidly to maintain their position in the water column. The maximum current should not exceed 50 to 100 cm/s.[293-296]

Net-pens should, in general, be sheltered from severe wind,[292] though some new designs for offshore net-pen facilities can handle a great deal more exposure to high wind and waves

FIGURE 20. Floats attached to the net-pen structure and to anchor lines keep modern net-pen systems in place and properly supported. Note the net stretched across the top of the net-pens (foreground) to protect the fish from bird predation. (Photo by Robert R. Stickney.)

than most systems that are currently in use. According to the literature, the depth of water over which the net-pens are located should provide at least 1 to 5 m of clearance between the bottom of the net and the substrate at low tide to allow for the dispersal of excess feed and fecal material.[293,294,297] In practice, greater distances between the bottom of the nets and the substrate may be desirable (see Section D below).

Suitable sites for net-pen facilities should have water temperatures ranging from 5 to 16°C, with optimum being 10 to 15°C.[293,294,297] Dissolved oxygen concentrations should be no less than 6 to 7 mg/l.[294-297]

Predation on fish in net-pens by birds and marine mammals has been a problem for many growers. Birds can be kept out of net-pens with netting stretched over the enclosures. External nets of large mesh and made of strong material which resists tearing by the teeth of marine mammals have been used as a deterrent around the net-pen structures. As a general principle, nonlethal control methods should be used against both bird and mammal predators.[298]

C. STOCKING AND GROWOUT

The bulk of the coho salmon produced in Washington have been aimed at the pan-sized market, where the fish compete well with rainbow trout cultured primarily in Idaho. The young salmon are maintained in freshwater hatcheries using rearing techniques outlined in Chapter 3 for approximately 18 months after which they are transferred to net-pens for an additional 7 to 12 months.[299] Investigators working in the Adriatic Sea, which is somewhat warmer than Puget Sound, have been able to produce 1,500-g fish in 8 months by stocking 40- to 80-g smolts in September.[300] Similarly, the Japanese are able to produce 3-kg coho salmon in 9 months from 150- to 200-g smolts.[4]

Harvest biomass of fish in Puget Sound net-pens is generally about 15 kg/m^3, whether the fish arc intended for the pan-size market or the so-called ''full-size'' market for which the fish run 2 to 5 kg.[292] British Columbia net-pen producers have targeted the full-size market with coho and chinook salmon in addition to that for pan-size fish. The numbers of pan-size salmon coming from Canada increased dramatically in 1989 and early 1990 when economically strapped farmers were reducing their losses by selling fish that might otherwise have been grown for the full-size market.

While most of the culture activity with Pacific salmon destined for the human food market is centered around coho and chinook, other species may be cultured in the future. There has not been much research conducted to date, though one study was undertaken in tanks with sockeye and pink salmon. Results of that work demonstrated that pan-size fish of 230 g could be produced from 4-g fingerlings in 280 days at 15°C in salinities of both 10 and 28 ‰.[301]

Net-pens have been utilized not only for producing fish for direct human consumption, but also for holding salmon prior to release. Delayed release programs, such as those which have been employed in Puget Sound, involve holding salmon smolts in seawater for a period of time and then releasing them into the wild. The additional time in captivity may enhance the percentage of fish that subsequently return to enter the recreational and commercial fisheries or escape upstream to spawn.[302] In Puget Sound, an 800-m^3 net-pen might be stocked with 20,000 fish averaging 20 to 25 g. Those fish will be released when they reach an average weight of about 75 g.[303]

D. ENVIRONMENTAL IMPACTS

The most likely adverse impacts arising from the presence of marine net-pens are in association with impaired water quality, particularly reduced dissolved oxygen and increased ammonia, and in the accumulation of waste feed and fecal material in the sediments, resulting in changes in the benthos community.[292,304] The fish within the net-pens will be among the first organisms impacted by impaired water quality. Problems can be avoided in well flushed areas. Proper rates of flushing will also limit the amount of organic material buildup under net-pens, but some accumulation may occur. In 1990, an Environmental Impact Statement (EIS) was published which dealt with net-pen sites in Puget Sound.[141] In terms of benthos monitoring, the EIS recommended that Washington state accept interim guidelines that were developed in 1986.[298] Among those guidelines were

- Maintenance of a minimum depth beneath the net-pens of 7 to 20 m (the minimum distance would be dependent upon current velocity).
- Habitats of special significance, for example, eelgrass beds, kelp beds, shellfish beds, and wildlife refuge areas, should not have net-pens located over them if the subject habitats are in water of 25 m or less, nor should net-pens be located within about 100 m of such areas in the direction of prevailing tidal currents or within 50 m in any other direction.
- Net-pens should not be located within 500 m of such habitats as seal and sea lion haulout areas, seabird nesting sites or colonies, or areas that are critical for feeding or migration of marine mammals or birds.
- To further protect marine organisms, including the fish within the net-pens, the antifouling agent tributyltin should not be used.

III. NUTRITION AND FEEDING

Pacific salmon have been fed in hatcheries for several decades, but it was not until the 1950s that semipurified diets specifically formulated to meet the needs of the fish were developed.[305] At about the same time, the Oregon Fish Commission and the Seafoods Laboratory of Oregon State University developed a practical diet, the Oregon Moist Pellet (OMP).[306] During the early days of net-pen culture, OMP (Table 8) was used as a standard

TABLE 8
Composition of the Oregon Moist Pellet Formulation Used in Pellets of 3.2 mm or Larger[30]

Ingredient	Percentage in diet
Herring meal (anchovy or hake meal may comprise up to one-half of the fish meal)	28.0
Wet fish	30.0
Fish oil	6.0—6.75
Wheat germ meal	remainder
Cottonseed meal (48.5% protein)	15.0
Dried whey product or dried whey	5.0
Corn distillers' dried solids	4.0
Trace mineral premix	0.1
Vitamin premix	1.5
Choline chloride (70%)	0.5

TABLE 9
Composition of an Open Formula Pelleted Dry Diet for Salmon[30]

Ingredient	Percentage in diet
Herring meal	50.0
Dried whey	5.0
Blood flour (or meal)	10.0
Condensed fish solubles	3.0
Poultry by-product meal	1.5
Wheat germ meal	5.0
Wheat middlings, mill run or shorts	12.22
Vitamin premix	1.5
Choline chloride (60%)	0.58
Ascorbic acid	0.1
Trace mineral mixture	0.1
Lignon sulfonate pellet binder	2.0
Fish oil or soybean lecithin (latter at maximum of 2.0%)	9.0

feed. However, advances in feed manufacturing technology with respect to dry extruded diets and the need to keep OMP frozen to prevent rapid spoilage made the diet economically impractical for growout operations. Thus, dry salmon diets, such as the one shown in Table 9, are commonly employed in salmon net-pens.

Research has shown that net-pen chinook salmon obtain little of their nutrition from planktonic or fouling organisms that attach to the nets. Feed supplied by the aquaculturist is the primary source of nutriton for fish in net-pens.[307]

The nutrient requirements of salmon have been determined only in part, with much of the work having been conducted on juvenile chinook salmon. Fewer studies have been conducted on the other species of Pacific salmon.[308] The results obtained in studies conducted to date have been broadly applied, and the nutritional requirements of Pacific salmon seem sufficiently similar to make broad application appropriate, though a good deal of fine tuning may occur in the future as additional studies are conducted on other species.

Salmon require relatively high levels of dietary protein and the amino acid requirements of the fish are best met with animal protein. For example, the lysine requirement of chinook salmon is 5.0% of dietary protein,[309] while that of chum salmon is 4.8%.[310] Chinook salmon

TABLE 10
Vitamin Requirements for Salmon[40]

Vitamin	Recommended level (per kg diet)
A	2,000 IU
D	2,400 IU
E	30 IU
K	10 IU
Ascorbic acid	100 mg
Thiamin	10 mg
Riboflavin	20 mg
Pyridoxine	10 mg
Pantothenic acid	40 mg
Niacin (nicotinic acid)	150 mg
Biotin	1 mg
Folacin (folic acid)	5 mg
B_{12}	0.02 mg
Choline	3,000 mg
Myoinositol (inositol)	400 mg

require methionine at 4.0% of dietary protein.[311] Those dietary levels cannot be met by plant proteins. Therefore, relatively high levels of fish meal are typically included in practical rations (Table 9) and attempts to substitute high levels of such plant proteins as soybean and cottonseed meal for fish meal have not been successful.[312] Plant proteins such as soybean meal can be utilized to a limited extent so long as the amino acid requirements of the fish are met.

Some studies on Pacific salmon have indicated that the fish have a dietary requirement for n-3 fatty acids. Coho salmon, for example, may require between 1 and 2% of dietary n-3 fatty acids.[313] Differences in the requirement for n-3 fatty acids may occur depending upon the nature of fatty acids present in the diet. For example, in chum salmon fry, 1% linolenic acid appears to be required, but the requirement can be met by about 0.5% of longer chain n-3 fatty acids.[314] The effects of dietary n-6 fatty acids on growth of Pacific salmon are mixed. There is some information from studies conducted with purified fatty acids which seems to indicate that a dietary level of linoleic acid of 1% enhances growth,[314] while levels above 1% inhibit growth.[313] However, higher levels of n-6 fatty acids present in practical diets have not led to reduced growth rates, at least in chinook salmon.[315] Clearly, more work on the fatty acids of Pacific salmon is required, though the present practice of incorporating fish oil into practical diets seems sound.

No carbohydrate requirement has been determined for salmon. Carbohydrates may have a protein sparing effect when present in salmon diets, and carbohydrates enhance pellet quality by serving as a binder.[316] In most formulations, carbohydrate levels are low.

Vitamin requirements for salmonid fishes have been established (Table 10), but there is relatively little information available on minerals. A limited amount of information on chinook and chum salmon requirements for minerals is available.[317-319]

Dry compressed pellets are typically fed to Pacific salmon in net-pens. The original formula was developed by the Abernathy Salmon Culture Technology Center[320] and has been modified over the years to reflect changes in our knowledge of the nutritional requirements of the fish.[306] Dry compressed pellets sink when thrown into the water. Floating pellets, produced by extrusion, can also be used but they are more expensive than compressed pellets. Because of the depths present and the active feeding habits of the fish, few pellets are lost from the net-pens. The water in Puget Sound net-pens is generally very clear, so observation of the fish is not difficult even when they do not approach the surface to feed. Thus, there is no real advantage in feeding floating pellets. Food conversion ratios (dry weight of feed offered divided by wet weight gain) are commonly 2:1.[292]

Various modifications in diets fed to brood fish have been made in an attempt to produce reproductive performance, but there has been little success except when the level of vitamins in the standard ration was doubled.[308] Some increases in fecundity and egg hatchability were observed when excess vitamins were fed, though the changes were not statistically significant.

Adding carotenoids to salmon diets has the effect of improving product quality by adding red pigment to the flesh, a trait which is commonly desired by the consumer. Carotenoid pigments are routinely added to fish diets in some countries but there are restrictions on their use in the United States. Good results have been obtained after 120 days by feeding 6 to 9 mg carotenoid per 100 g of feed to coho salmon weighing over 215 g.[321] The subject has recently been reviewed.[322]

Steroid hormones have been added to salmon and trout diets to induce sex reversal or enhance growth. The hormone 17β-estradiol has been used to sex-reverse male Atlantic salmon and rainbow trout.[323] Inadequate exposure of the fish to the steroid offered in the food of the fish led to the development of hermaphrodites. Various steroids have been used in attempts to enhance growth in salmonids. Examples are bovine growth hormone, 17α-methyltestosterone, 11-ketotestosterone, testosterone, 4-chlorotestosterone acetate, oxymetholone, progesterone, estradiol, dimethazine, norethandrolone, and L-thyroxine.[324-329]

n has been produced in coho salmon by 17α-methyltestosterone, stosterone, oxymetholone, and estradiol.[325,328] Gains in length and ement in condition factor, were reported for both chinook and coho ltestosterone in one study.[324] While there may be some growth ad- ormone in fish diets, such chemicals have not been approved for that not being utilized in the United States by net-pen producers of Pacific

IV. SPAWNING IN SALTWATER

m the use of wild or public hatchery fish as brood stock is not only en farming industry, it is often necessary because of regulations which n from the public domain in commercial aquaculture. Also, many net- he desire to develop their own selected lines of fish in an effort to erformance under the rather unique rearing conditions that are associated ion. Producers often have their own hatcheries and depend upon pro- is a source of eggs for those hatcheries.

tion of the fish produced in a net-pen facility will be reared to adulthood. o mature, they will moved to freshwater for spawning, though spawning sometimes desirable when onshore spawning facilities are limited. In ted with ocean ranched coho salmon in Oregon, returning fish were l in either marine net-pens or freshwater to mature.[330,331] Survival and duced in fish held in the marine environment, perhaps because the fish motic stress. In general, attempts to maintain coho salmon in full strength wning have been unsuccessful.[332]

emonstrated that coho salmon can be effectively spawned out of seawater a distinct halocline present which allows the fish to select a suitable n no halocline is present, a low salinity lens can be created by pumping he surface of the net-pen.[4]

TABLE 11
Therapeutants Registered and Approved for Use in Foodfish Aquaculture by the U.S. Food and Drug Administration or the Environmental Protection Agency[65]

Chemical	Use pattern
Formalin	Parasiticide and fungicide for salmonids
Oxytetracycline (terramycin)	Antibacterial for salmonids
Sulfadimethoxine and ormetoprim (Romet-30, Romet-B)	Antibacterial for salmonids

V. DISEASES AND THEIR CONTROL

In Washington, fish farmers lose, on average, about 50% of their stock between egg fertilization and marketing of the final product.[335] In addition to losses attributable to predation, the physiological stress associated with transfer from freshwater to seawater at smolting, escapement from net-pens, handling stress, and environmental problems such as noxious algae blooms (discussed in the following section), diseases can be an important source of mortality. When such losses go unaccounted, they can lead to increased food costs,[336] and, of course, have negative impact on the economic viability of the fish farming operation.

Salmonids, like other fishes, are susceptible to a variety of diseases, including viruses, bacteria, fungi, and various types of parasites.[337,338] Losses of fish can carry over into saltwater rearing, as is the case with bacterial kidney disease. Vibriosis, attributable to the bacterium *Vibrio anguillarum,* is a common problem in net-pens.[45] The disease, which tends to occur during warm periods, but can occur anytime, is characterized by the presence of lethargic fish and those which swim erratically. Diseased fish will show hemorrhaging from the bases of the pectoral and anal fins and may have a bloody discharge from the vent. Spleen enlargement is typical. Crowding and handling stress are often associated with onset of the disease.

Concern has been expressed in some quarters that diseases which proliferate among cultured fish can result in epizootics within wild fish populations. In reality, the transmission of diseases from wild to captive fishes would seem to be far more likely than the reverse.[292,304] Salmon can be successfully vaccinated against vibriosis,[339] and routine vaccination of fish before transfer to seawater is common in the net-pen farming industry.

The number of therapeutant chemicals approved for use by fish culturists in the United States (Table 11) is limited.[340] Opponents to net-pen culture in Washington have suggested that the use of the antibiotic oxytetracycline could lead to the development of resistant strains of bacteria in the marine environment. Examination of that potential led Weston[292] to the conclusion that because of the high water solubility of the compound, its rapid rate of dilution into the water, and the infrequency of use, potential impacts are minimal. Net-pen fish culturists only utilize the antibiotic when they are convinced that an epizootic is imminent or when one has been confirmed and they feed the drug for only 10 days.

VI. PLANKTON BLOOMS

Blooms of noxious phytoplankton have caused significant problems throughout the net-pen salmon industry, both in North America and Europe. For example, during 1987, at least 250,000 Atlantic and Pacific salmon of various ages were lost by private producers in the state of Washington to noxious algae blooms. The monetary loss was placed at over $0.5 million.[341] In British Columbia, algae blooms killed up to 3,500 tons of farmed salmon in 1989, causing a significant carcass disposal problem in addition to severe economic loss.[342]

The actual causes of mortality in association with algae blooms have not been precisely determined; however, mechanical injury to gill tissues, direct toxicity from chemicals produced by the algae, and physiological imbalances have all been hypothesized as causitive factors. Mechanical injury or actual clogging of the gills by phytoplankton cells can lead to excess mucus production and possible suffocation.[343-345]

The conditions required for development of phytoplankton blooms in marine waters are well known. Included are relatively stable surface waters, an adequate amount of sunlight, sufficient levels of required nutrients, and the presence of phytoplankton cells which can form the basis of the bloom.[340] What is not known is how to predict the development of a bloom of a species that might be toxic to fish in net-pens.

Among the species which have been associated with noxious blooms of phytoplankton in the Pacific Northwest is the diatom *Chaetoceros consulutus*. That species has been shown to kill sockeye and coho salmon smolts in British Columbia when the fish are exposed to levels of several million cells per liter over one to seven hours.[343] The concentration of cells required to kill 50% of exposed small chinook and chum salmon, assuming algal chain lengths of 3 to 10 cells each, ranged from 3 to 18 million cells per liter in a study conducted in Alaska.[346] Anecdotal evidence suggests that coho and chinook salmon are more susceptible than Atlantic salmon to *Chaetoceros* blooms.[341]

The dinoflagellates *Ceratium fusus* and *Gymnodinium splendens* have been found in association with mortalities of both pen-reared salmon and pandalid shrimp in Puget Sound.[347] Microflagellates may also be a problem for net-pen salmon producers. *Heterosigma akashiwo* has been implicated in salmon losses in British Columbia and may have been the cause of a fish kill in Washington during 1986.[345]

In the summer of 1990, reports in the news media indicated that a toxic algae bloom in Puget Sound, attributed to *Heterosigma* spp., was responsible for salmon losses into the millions of dollars. Salmon net-pens were towed from their normal mooring places into temporary locations that did not exhibit high concentrations of algae. The algae were apparently present in high concentrations only in the upper portion of the water column, so fish in net-pens were particularly susceptible to the bloom while free-living populations seemed unaffected. One fish farm found concentrations of the organism at one million cells per liter,[348] while the net-pens associated with the National Marine Fisheries Service at Manchester, Washington may have developed algal cell densities in excess of four million cells per liter.[4] In the latter instance, a captive brood stock which was being maintained as a source of gametes for the threatened White River spring chinook salmon stock was virtually eliminated.

Concerns have been expressed in some quarters that nutrients being added to net-pens by way of the feed provided to the fish stimulate the development of noxious phytoplankton blooms. Examination of the relative amounts of nutrients provided to the water from feeding net-pen fish and those entering from other sources have led to the conclusion that activities associated with salmon farming have not and will not cause or intensify a phytoplankton bloom in the main channels of Puget Sound.[288,349]

Chapter 5

CULTURE OF ATLANTIC SALMON

Arni Isaksson

TABLE OF CONTENTS

I. INTRODUCTION

During the past 20 years there has been a great increase in the farming of fish and other aquatic animals. This is partly due to the fact that most wild fish stocks are being harvested at or near maximum capacity or have been overharvested. Aquaculture production is expanding most rapidly in the developing countries, where emphasis is either on the use of inexpensive methods to produce fish protein to feed a rapidly growing human population or on the production of luxury products which enter the international markets and provide foreign exchange for the producer nation (e.g., shrimp culture).

In the industrial countries highly technical methods have received major emphasis. The primary goal has been to supply international fish markets with relatively high-priced aquaculture products. Large-scale production is limited to relatively few species; primarily salmon, trout, catfish, sturgeon, shrimp, and several types of molluscs. Culture of salmonids can be broken into several categories depending on the species and rearing methods used.[350]

Enhancement is a term used for the efforts of increasing the runs of salmon to a particular river or river-system by releasing fry or smolts into the system. Other activities such as the construction of fishways and other river improvement activities may also be included. Since the fish are spending most of their lives in natural waters, this form of culture is often called extensive aquaculture. In a typical extensive culture system, however, the fish are not free to leave an enclosed area.

Another form similar to extensive aquaculture is salmon ranching wherein the smolts are released directly into natural waters, where they are free to roam long distances and grow to market size on natural foods. The feeding area can either be a lake or the ocean. Harvest of the salmon in many cases depends on their homing instinct as they are caught close to the home stream on their spawning migration. In other instances the harvest may take place in the feeding areas as in the case of the Baltic Sea salmon fisheries. Ranching has gained its highest level of economic importance with respect to the various salmon species which occur on both sides of the Pacific Ocean.

The most recent development in salmonid culture is intensive culture. In this method the fish farmer takes the fish through all stages from egg to adult, using modern rearing techniques and artificial diets. A good example is the salmon farming industry in Norway. Since the fish are fed a high protein diet, the end product is usually expensive.

This chapter primarily discusses the aquaculture of Atlantic salmon as it relates to the production of food fish. The first section deals with the historical development of rearing and ranching. Then there are two main sections dealing with freshwater and saltwater rearing with emphasis on technology, rearing practices, and various rearing processes. The next sections deal with biological problems in rearing as well as environmental concerns, both those threatening rearing facilities and those resulting from the operation of those facilities. The final sections deal with the development of rearing and ranching in various countries, production figures, and current economics.

II. HISTORICAL BACKGROUND

Artificial fertilization methods were first discovered by Jakobi,[351] but it took many years before propagation of Atlantic salmon (*Salmo salar*) became practical. European hatcheries started planting sac fry into streams in the early 1800s, but only with limited success until methods for feeding the salmon fry were developed.

The initial success in the rearing of salmonids in Europe was actually associated with domesticated rainbow trout (*Oncorhynchus mykiss,* formerly *S. gairdneri*)[352] in Denmark. The species was imported to Denmark from the west coast of North America in the late

1800s. The rearing program initially developed slowly, but by 1950 the production was about 2,000 tons of portion-sized (pan-sized) fish.[353] By 1975 it was over 15,000 tons, where it has stabilized.

The farming of rainbow trout has since expanded in many other European countries, where the climate is favorable for that activity. In Norway, where the climate was not particularly suited for the freshwater rearing of rainbow trout, the growing of portion-sized trout never was competitive, but the Norwegian producers started producing larger foodfish, in many cases in seawater enclosures.

Rearing of Atlantic salmon to parr size began in Sweden around 1930 and Atlantic salmon smolts were first produced in Sweden about 1950 under the leadership of Dr. Börje Carlin of the Swedish Salmon Research Institute.[354]

The major breakthroughs in the rearing of Atlantic salmon came with the development of dry feeds in the early 1960s. Those were largely based on the pioneering work of J. E. Halver and co-workers relating to protein and vitamin requirements of Pacific salmon. Dry feeds for Atlantic salmon were initially developed by Astra Ewos in Sweden for smolt rearing but were soon developed for the production of rainbow trout and salmon foodfish by many feed companies.

With the advent of smolt rearing, salmon enhancement and ranching started increasing in several countries. In 1960 the Swedish salmon rearing stations were producing over half a million smolts, and by 1985 the total smolt releases into the Baltic Sea by Sweden and Finland were over 5 million smolts.[355] In the early 1960s two Atlantic salmon research stations were built in Ireland and Iceland. The Irish station, located at Furnace on the Burrishole river and operated by the Salmon Research Trust,[356] has contributed greatly to the understanding of salmon ranching principles. Kollafjördur Experimental Fish Farm in Iceland, which has a large share of its income from the sale of returning salmon, must be considered the pioneer of private salmon ranching with Atlantic salmon.[350] Subsequently, similar experimental stations were built in St. Andrews, Canada,[357] at Air in the Faroes,[358] at Ims in Norway,[359] and at Lussa in Scotland.[360]

Commercial salmon farming in Norway is considered to have started with the experiments of Ivar Heggen and the Vik brothers in the 1950s.[361] These forerunners built an industry by trial and error. The Vik brothers discovered that rainbow trout could live in full seawater after a short period of acclimation from freshwater. They also showed that spawned-out wild salmon would take food in captivity. In 1960 the brothers started a large land-based seawater rearing facility with emphasis on rainbow trout culture. Many others started farming trout in net-pens. Most, however, did not think that the technique would work for Atlantic salmon.

In the mid 1960s, Sivert and Ove Gröntvedt started rearing Atlantic salmon in net-pens from smolts. This process worked well and now over 95% of the farmed Atlantic salmon are raised this way. By 1970 the production had reached 4,000 metric tons[362] and a major industry had been born.

III. TECHNICAL DEVELOPMENT

Salmon farming has evolved into a fairly technologically advanced industry over the past 40 years. There has been a shift from earthen ponds to concrete or fiberglass tanks and net-pens, and dry feeds have almost completely replaced wet feeds. Computers are being used to calculate feed conversion ratios and to control feeding. Recirculating units are getting more common to save energy and water. Most of the technology is useful but can in some cases turn into a fish farmer's nightmare, particularly if it becomes too complicated.

The most common equipment used in the hatching and rearing of Atlantic salmon in various countries is described in this section. Although the principles are fairly standard, the equipment used varies considerably depending on locality, environmental conditions, and various other factors.

A. HATCHING AND REARING FACILITIES

1. Hatcheries

Due to the light sensitivity of the eggs, the hatching of Atlantic salmon always takes place indoors where amount of light can be controlled. The oldest method for Atlantic salmon hatching employed "California" type hatching troughs, developed for rainbow trout and Pacific salmon. These were originally made from wood, but modern troughs are made from fiberglass with four to seven hatching trays in each unit. Water flows longitudinally along the trough and wells up through the bottom of each hatching tray. Hatching troughs are still the most common method of hatching Atlantic salmon eggs. The hatching troughs are often placed in parallel rows of two and stacked two or more levels high. In many cases the fry stay in the hatching trays until they are ready for feeding; sometimes an artificial substrate is provided to minimize movement and energy expenditure.[363]

Various types of incubating cabinets are also used in some stations. They save on space and water, but egg inspection is more difficult when they are employed. Hatching cylinders have also been used, in which eggs can be processed up to the eyed stage in large numbers. In this case dead eggs can not be picked and one must rely on disinfection to combat problems with the development of fungus.

2. Smolt Stations

Startfeeding and early rearing of Atlantic salmon usually take place indoors in relatively subdued light. In the early years of rearing, the salmon fry were startfed in hatching troughs after the egg trays had been removed. The technique was commonly employed for rearing rainbow trout in Denmark, after which the fish were put into large earthen ponds for growout to portion size. Initially, Swedish growers thought that earthen ponds would be suited for overwintering of salmon smolts and a few stations started operating on that principle. The approach was, however, soon abandoned in favor of concrete or fiberglass tanks (Figure 21).

In the 1960s Swedish salmon hatcheries started using square 2- to 4-meter fiberglass rearing tanks with central drains. Swedish salmon smolts, which require 2 years to produce, were kept in the tanks for the first year of rearing and then put into large circular concrete tanks with central drains, where they were kept over the second winter (Figure 22). The concrete tanks usually had electric heaters in the walls to prevent damage from ice formation associated with the low incoming river water temperature and harsh climate.

As salmon smolt production facilities proliferated, the Swedish rearing method was adopted in various other countries. The Swedes became major suppliers of rearing equipment such as tanks, feeders, and fish-pumps. They also developed and distributed feeds. With the growth of the Norwegian industry, many more companies in Norway and Denmark got involved in equipment and feed manufacture.

The Swedish type tanks are either rectangular or round, but are all based on the principle of circular flow and a central drain, with an internal or external standpipe. Rectangular fiberglass tanks ranging from 1 to 4 m^2 are most commonly used for early rearing, sometimes in racks of two or more levels.

Large circular tanks (over 5 m in diameter) are often used to hold fish that have reached smolt size. The tanks are frequently made with fiberglass walls and a smooth concrete bottom, but tanks constructed entirely of concrete are also common (Figure 23). In Scotland, a number of farms have used corrugated iron wall sections to fabricate circular ponds. In

FIGURE 21. Swedish smolt rearing stations led in the early technology of smolt production. Square fiberglass tanks are most common for initial rearing in the large smolt farms operated by the Swedish Electrical Board in their large salmon enhancement programs. (Photo by Arni Isaksson.)

most countries large tanks are placed outside, but in a colder climate such as that found in Iceland, where drift-snow and frost are a problem, they are often under a roof.

3. Net-Pens

As Sweden took the lead in smolt rearing, Norway has been the pioneer in the production of adult salmon. Initial trials were conducted in land-based units, but the bulk of the production now takes place in net-pens (Figure 24). The same holds true if one considers the rearing of Atlantic salmon on an international basis. Land-based facilities are still experimental and marginal in economic terms.

Net-pens can basically be divided into four types:[364] fixed, floating, submersible, and submerged. Of these, the only one of real importance in salmon culture is the floating type. Some fixed nets with poles driven into the sea-bottom have been successfully used in Norway and some of the more recent ocean pens are submersible. Submerged pens are of no present importance in commercial salmon production although a number of designs have been proposed.

Experience has shown that pens are by far the most ecomomical way of rearing salmonids commercially.[364] Pens, however, have the disadvantage of being relatively vulnerable to a number of problems. They are often situated in public or multi-use water bodies and may be exposed to pollution as well as to losses caused by predators or vandalism. They are also more susceptible to storm and ice damage than most other types of rearing systems.

Salmon pens are basically a sturdy frame that floats on the sea, sometimes with a built-in working platform. Frequently, an inside railing is provided for the safety of rearing

FIGURE 22. Large concrete outdoor tanks, such as these at Green Lake hatchery in Maine, are most common for storage of smolts during the last few months before release both in Europe and North America. (Photo by Arni Isaksson.)

personnel and to prevent the fish from jumping out of the net-pen. Hanging from the main frame is a net enclosure, frequently over 4 m in depth, where the salmon are contained and fed. In the sheltered Norwegian situation with deep fjords and a small tidal range, the pens are often attached to a pier which connects to the land. In other countries, such as Scotland, Ireland, and Iceland, with shallower fjords and a greater tidal difference, net-pens are frequently placed a considerable distance from the shore and have to be serviced by boats.

Many types of net-pens have been developed. Only some of the more typical ones are discussed here. The early net-pens were octagonal in shape and comprised of eight wooden units connected by rubber joints on the corners.[365] Each unit was 5 m long, making a total circumference of 40 m. Flotation was usually provided with styrofoam floats. The pens were frequently 12 m in diameter with approximately 500 m^3 rearing capacity. A fairly convenient working platform was built into the frame. These structures were fairly well suited for the sheltered Norwegian fjords, but could not withstand the wave action in more exposed areas as in Scotland, Ireland, and Iceland.

Some commercially available pens have been designed to hold the netting afloat. Such pens basically float on rubber buoys without any rigid framework at the water surface and are thus very flexible.

A more recent development is a circular frame made out of buoyant PVC (polyvinyl chloride) tubing, with an attached railing which goes under the trade name ''Polarcircel''. These pens are commercially available and can withstand more wave action than the hexagonal pens described above. Sometimes a working platform is created supported by double tubing, but more commonly the pens are attached to a pier that connects the net-pens with the land.

FIGURE 23. Norwegian smolt farms, being in a milder climate than in Sweden, are less capital intensive. Large circular tanks with a fiberglass lid have been used for the rearing of smolts from the swim-up stage through smoltification. Start-feeding is, however, more commonly performed indoors in smaller (2 × 2 m) tanks. (Photo by Arni Isaksson.)

Many sturdy net-pens are available commercially, some of which are connected to extensive working platforms. These pens, which are stronger and more expensive than the more traditional pens developed in Norway, have been most commonly used in Scotland, Ireland, the Faroes, and Iceland, where the coastline is more exposed and winds more prominent. In Scotland, the most common pens are square in shape, 8 × 8 m, and are of wooden construction.[366] These pens have a rearing volume of 250 m^3, which is only half of the volume of the average Norwegian pen. Under the severe Scottish weather conditions one can expect to lose a small fraction of these pens in severe storms.

Oceanic pens have also been extensively used in Ireland, Scotland, the Faroes, and Iceland. One of the more common oceanic pens is "Wavemaster", a square steel construction, 12 × 12 m, used extensively in Ireland, Scotland, and Canada.[367] Another common type is the "Bridgestone cage", which is polygonal and composed of flexible rubber tubing with steel corners (Figure 25). It has a rearing volume of 10,000 m^3 and must be serviced by a large specially designed boat. The "Farm Ocean", another oceanic giant, is of semisubmersible steel construction and variable buoyancy. It has a feed silo and a built-in feeding mechanism for dispersal of dry feed and can operate without human attention for a few days in the case of bad weather. Another oceanic net-pen is the "Ewos giant cage", which is practically a floating fish farm with feed storage and a fish-processing area. Many other designs of oceanic pens are being used in various countries bordering the Atlantic Ocean.

4. Sea-Enclosures

Sea-enclosures have primarily been used in Norway. They can be defined as fixed net-pens or impoundments of sea water, where most of the circumference is formed by a natural

FIGURE 24. Circular cages of PVC plastic tubing are commonly used by the salmon farming industry in Iceland. The pictured cages, located in Hraunsfjördur on Iceland's west coast are used for final rearing and release of smolts in an ocean ranching operation. Total releases in ranching now exceed five million annually. (Photo by Rafn Hafnfjord.)

shoreline. Man-made barriers on one or more sides of the enclosure, which permit passage of water, complete the enclosure.[353] These barriers can, in their simplest form, be made out of nylon netting suspended from pilings, although their use is relatively dependent upon weather and local topography. The only successful system has been operating in Norway.[353] Much more elaborate concrete and steel structures were built by A/s Mowi in the Bergen, Norway area about 1970. The company was the largest single producer of salmon in Norway, producing over 500 tons in two enclosures with a combined rearing volume of 230,000 m^3. The larger enclosure, however, comprised over 80% of the rearing volume. It soon became apparent that tidal exchange would not be satisfactory to ensure proper circulation of water in the enclosures, especially in the deepest part. The company solved that problem by installing electric pumps which sucked water and sludge from the deepest part and discharged it outside the enclosures. Additional pumps were used to increase flow through the enclosures.[353] The company operated the large enclosure until 1987, when it was replaced by a matrix of 40, 15×15 m^2 net-pens with extensive walkways.[368] Having a rearing volume of 50 to 80,000 m^3, it is now the largest net-pen rearing operation in the world.

5. Land-Based Operations

In Norway, land-based operations were built in the late 1950s. These were made of concrete and were based on continuous pumping of seawater. Most of these units are no longer used for the rearing of food fish but rather for production and acclimatation of smolts.[369] In Scotland, where natural conditions for net-pen farming are less favorable than in Norway, a number of land-based operations were established in the late 1970s. The same was true for Iceland in the mid-1980s.

FIGURE 25. The Bridgestone oceanic cages are in operation in several countries where shelter is limited, e.g. Ireland and the Faroe Islands. The flexible rubber cage with a rearing volume of 10,000 m^3 is difficult to tend by conventional means. Special tender boats have been designed for that purpose. (Photo by Rafn Hafnfjord.)

The primary advantage of tanks on land compared with pens in the marine environment is the improved control of the fish culture environment. In some cases the fish farmer can control temperature as well as salinity and dissolved oxygen. Flow rates can be controlled and oxygen added to provide optimum growing conditions. There is very little risk of losing the fish because of foul weather or predation. Grading as well as disease treatment and slaughtering are also easier to perform.

Land-based farms do, however, have their drawbacks. The construction and operation costs are much higher than those associated with pens. Accidents due to electrical or mechanical failures are common in spite of expensive back-up electrical equipment. Costly alarm systems are essential.

Land-based operations are most popular in Scotland and Iceland. A variety of designs are currently in use, although circular tanks with a center drain seem to be a preferred design.

The Scottish operations have in some cases been using concrete raceways or large earth raceways lined with polyethylene. The most common design, however, employs large circular ponds with central drains and an external standpipe, similar to those used in smolt rearing. These are commonly 12 m in diameter with a total depth of 2 to 4 m. A few large units are up to 25 m in diameter. Most of the larger pools are constructed from concrete, but smaller ponds sometimes have walls made out of several sections of corrugated iron or fiberglass sheets bolted together.

In Iceland, several large land-based farms have been built with a combined rearing

FIGURE 26. Land-based operations using geothermal heat to enhance the rearing process are fairly common in Iceland, including this 600-ton facility, Islandslax, located in southwestern Iceland. (Photo by Rafn Hafnfjord.)

capacity of over 1,000 tons (Figure 26). These facilities all feature circular tanks, most of them constructed from concrete with diameters of 12 to 25 m and total depth of about 4 m. Due to heavy wave action at some of the sites, the facilities can not pump water directly out of the sea and have had to drill seawater wells on the beach. In some cases the stations have had to contend with ambient annual temperature variations in their well water, but in one instance the wells supply seawater with a constant temperature of 7°C year round. That temperature is at the low end of the range for rearing adult Atlantic salmon, but the company has not had the financial capacity to invest in a heating system.

The original plans for land-based rearing in Iceland assumed that such facilities would be combined with steam power plants in suitable areas and that they would utilize waste heat. The connection with power plants has not materialized, but the salmon farms were built mostly in the late 1980s.

Due to the unstable weather and electricity in Iceland, the land-based stations have all had to invest in costly diesel powered generators. These have been fairly dependable, but add greatly to the immense construction cost of these stations.

B. HEATING AND RECIRCULATION TECHNIQUES

Heating of rearing water is most commonly employed in conjunction with the rearing of juveniles where relatively small water quantities are involved. Most commonly, heat exchangers, heat pumps, or electric heaters are used. Heat exchangers transfer heat from one water source to another through metal plates or tubes. They are used extensively to heat freshwater with geothermal water in Icelandic smolt stations and to gather waste heat from industrial effluents in many countries.

Heat pumps give off more heat energy than they get from the electricity which drives them. This is possible because the heat pump extracts heat energy from a water source and adds it to the energy from the electric heating process. Heat pumps are used in several smolt farms in Europe.

Direct heating of water for salmon culture with electricity is only practiced when very small quantities of water are involved, such as in conjunction with the hatching of eggs. Heating water for egg hatching alone is not desirable as it produces fry which are out of phase with the ambient freshwater temperatures commonly used for startfeeding and rearing.

In Norway, where freshwater temperatures are very low in the winter (0 to 2°C), the rearing water for parr has been heated by pumping seawater of 6 to 8°C from deep fjords and mixing it with freshwater. The growing period can thus be extended and the proportion of one-year smolts produced increased considerably.[353]

Recirculation techniques have been used extensively in some rearing stations for Pacific salmon to conserve water and energy. In most of those stations 90% of the water flow is reused, and only 10% is new water. The technology for recirculation of water for Atlantic salmon production is, in theory, already at hand, but no commercial stations have been constructed which depend entirely on such systems, as they are considered too experimental for commercial growout.[370] Some smolt stations are, however, using recirculation temporarily for hatching and startfeeding.

C. FEEDING TECHNOLOGY

1. Dietary Requirements

Diets of animals in captivity should ideally satisfy all nutritional requirements. Since fish are often fed on a single diet, any imbalance or deficiency in the diet may be quickly reflected in fish health or performance. In the 1940s several researchers found that vitamin deficiencies in diets were responsible for or augmented the symptoms associated with certain types of pathogens.[371] In the late 1950s, J. E. Halver and co-workers established the protein requirements of chinook salmon, which laid the foundation for many subsequent studies on dietary requirements of other fish species.[372]

2. Diet Preparation

During the 1920s and 1930s, salmon and trout were fed a variety of feeds based on the locally available ingredients. These ingredients included salmon eggs and fresh, canned, or frozen fish, combined with various animal components such as spleen and liver.[306] In the 1940s demand for ingredients used in wet fish feeds increased due to increased hatchery production and other applications. To extend the traditional ingredients, meat-meal mixtures for salmon were developed. Several feed mixtures were made available commercially.

Meat-meal mixtures are still used in salmonid aquaculture. In Europe, fish-processing waste and fish silage are combined with dry meal in various proportions to form wet feed. Semi-moist diets were then developed, including the Oregon Moist Pellet (OMP), which is used mostly for rearing Pacific salmon. Such diets have also been tried in conjunction with Atlantic salmon production in Europe.

Dry compressed pellets became available in the late 1960s and the formulations of Phillips et al.[373] and Fowler and Burrows[320] provided the foundation for commercial production of feeds used throughout the world.[306] Many of the diets are open formula and are continually modified and tested in various laboratories. Dry compressed pellets are most extensively used in smolt production in various countries and enjoy an increasing share of the feed used for the production of food fish.

3. Feeding

Hatchery managers desire to grow with minimum feed wastage. Dry feed manufacturers have developed feeding guides for various diets and species of fish. The amount fed and feeding frequency varies with temperature and size of the fish. More frequent feeding is desirable with smaller fish, in particular during startfeeding. Particle size is very important. If the feed is too large the fish can not consume the pellets. Similarly, if the feed is too small it may not be consumed. In either case, feed is wasted and may have a negative impact on water quality, and the fish will not grow.

Atlantic salmon fry are normally started on a fine mash, but as they grow the size of the pellets is increased. The sizes range from <0.6-mm startfeed to 3-mm pellets when the parr are around 20 grams. The pellets used for larger fish range from 3 to 4 mm for smolts to 6 to 7mm for brood fish.[374]

Dry feeds have been enjoying an increasing share of the market in recent years. In excess of 85% of the salmon reared in Norway receive dry feeds.[362] Such feeds are more convenient than the wet feeds and give better growth at sea temperatures above 5°C.[375] At lower temperatures there is some evidence that wet feeds are superior. Dry feed research is, however, quickly reducing this difference.

Dry feeds are usually fed with automatic feeders of various types. Some feeders, suspended over rearing chambers, use electric motors and varying sizes of plates to distribute the right amount of feed. Others use pneumatic devices to throw the feed over the surface of the rearing container. Automatic feeders are usually used in conjunction with computers, clocks, or other timing devices which can be set to provide the desired feeding schedule.

Wet and semi-moist feeds are still commonly used in the feeding of Atlantic salmon to market size. Moist feeds are usually made from chopped fish, often frozen capelin, mixed with a dry meal formula containing binders, vitamins, and minerals.[374] Some fish farmers also substitute a part or all of the chopped fish with liquefied products such as acid-preserved, hydrolyzed fish-processing waste.[376] In most cases the proportion of wet material in the diet is about 60%. Moist and wet diets are usually fed by hand although some automatic feeding methods have been developed and moist diets are sometimes fed with pneumatic feeders.

Flesh color is of major concern in the marketing of Atlantic salmon. Crustacean waste, which contains the carotenoid pigment astaxanthin, is sometimes mixed with wet and moist feeds to produce the desired red coloration. The commercially synthesized compounds canthaxanthin and anthaxanthin are commonly used in dry feeds to enhance flesh color. These feeds, which contain 4 to 6 mg carotenoid per 100 g of feed, must be fed for at least 4 to 6 months before slaughtering to provide acceptable color.[377]

IV. HATCHING AND FRESHWATER REARING

Hatching and rearing methods for Atlantic salmon are basically similar in most countries, although sizes and shapes of rearing tanks may vary for fry at various stages. It should also be pointed out that intraspecific differences in fish found among various countries may vary considerably with respect to some biological aspects such as optimum hatching and rearing temperatures. An Atlantic salmon in Canada can thus withstand temperatures several degrees higher than an Icelandic salmon due to adaptive and evolutionary processes. It is sometimes difficult and even dangerous to transfer information on the same species directly from one country to another. This section is based primarily on Norwegian and Icelandic information, which appears to be fairly typical for northern Europe.

A. PHYSICAL AND BIOLOGICAL PARAMETERS

The main parameters of importance in a hatching and rearing operation are location, water quality and flow, temperature, fish densities at various stages, and mortalities.

1. Water Quality

Atlantic salmon are hatched and reared in freshwater and move as smolts to sea, where they grow to adult size (see Chapter 1 for additional life history details). Culturists normally only rear Atlantic salmon to smolt size in freshwater. Subsequent rearing takes place in saltwater. Icelandic experience, however, has shown that adults can be easily grown in low salinity lagoons and even in freshwater.

Young salmon are fairly tolerant of turbidity, temperature variation, and other variable water quality conditions that they would normally encounter in a freshwater stream. The eggs, however, which normally are deposited in the bottom gravel of rivers, are very sensitive to siltation and need high quality water. The fry are also fairly sensitive to water impurities during startfeeding and early rearing. Both eggs and young fry are sensitive to various metals, such as copper and zinc. Plastic plumbing is recommended for smolt production facilities as a means of reducing exposure of the fish to metals.

The source of water used for rearing varies geographically. In Norway, where freshwater wells are scarce, most hatcheries and smolt stations use runoff water or water from freshwater lakes. Various other countries also use lake water. Water quality can be quite satisfactory, but the source normally contains fish and various pathogens, which can be easily introduced into the culture facility. Some stations have installed costly filtration and sterilization equipment to avoid the problem. The use of river or lake water is not feasible in some areas of Northern Europe and North America, due to acidity problems often associated with acid rain.

Iceland, being of recent volcanic origin, has a number of freshwater wells both cold (3 to 5°C) and of geothermal origin. Due to heavy rainfall and the relative absence of trees, the runoff streams tend to be exceedingly turbid during heavy rainfalls. Smolt production stations have, as a consequence, turned entirely to using heated well water for their hatching and rearing operations. The freshwater wells, in addition to being very clean, are usually devoid of fish and fish pathogens, which eliminates transmission of diseases by way of the water source. The water, however, needs to be heated from 4 to 12°C to be suitable for fish rearing.

2. Hatching and Rearing Temperatures

In nature, hatching of Atlantic salmon takes place in fairly cold stream water during the winter. The time of hatching depends on location and the ambient water temperature in the stream, with the process being accelerated in warmer streams. It has been established that Atlantic salmon eggs take approximately 500 degree days (average temperature × number of days) from fertilization to hatching. This means that the process takes 100 days at an average temperature of 5.0°C and 50 days at an average temperature of 10°C.

In most countries which use ambient water temperatures for rearing, there is no reason to speed up hatching as ambient temperatures must be over 8°C for successful startfeeding.[378] In countries like Iceland, which has complete control of hatching and rearing temperatures, it is most common to speed up hatching and startfeeding by several months. This is also true for several smolt stations in Norway, which use cooling water from hydroelectric power plants.[353]

Figure 27 compares several hatching programs used at Kollafjördur Experimental Fish Farm in Iceland. Enhanced hatching cycles are common and are prerequisites for successful one-year smolt production. Experience has shown that the optimum temperature for startfeeding and rearing of Icelandic stocks is 12 to 13°C. In North America temperatures of 12 to 15°C seem to be best.[380]

In Iceland, rearing temperature is fairly constant at 13°C until the fry reach smolt size (20 to 30 g) at which point the smolts are placed in outdoor ponds for overwintering in

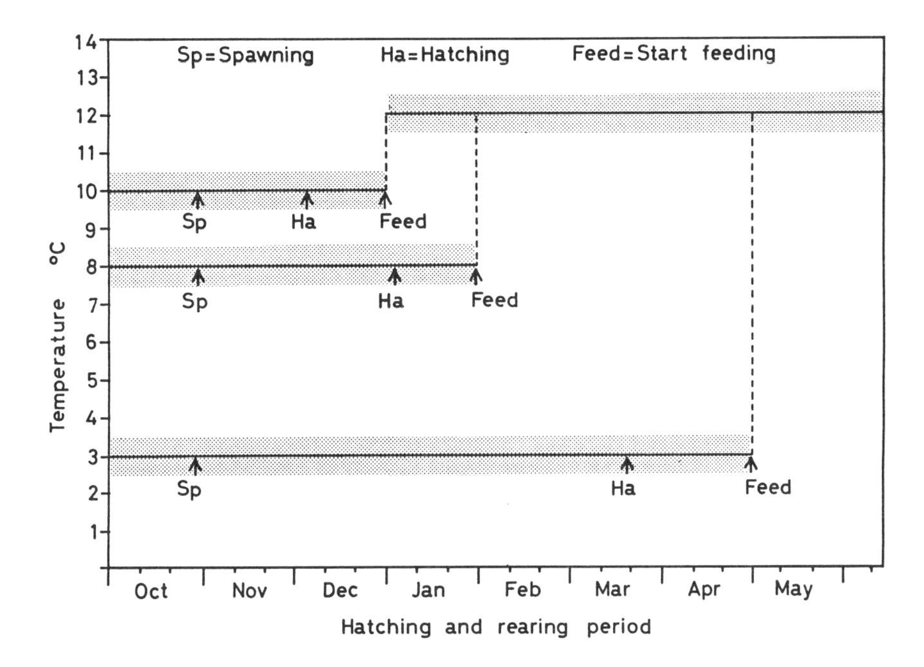

FIGURE 27. Enhanced hatching programs are prerequisites for successful production of 1-year smolts of Atlantic salmon. The figure shows the effect of water temperature on incubation and time of startfeeding in Icelandic salmon culture.[379]

constant 3 to 4°C well water. In other countries, where ambient water is used, higher summer rearing temperatures are common (15 to 17°C) and overwintering is often in water temperatures down to zero. In those stations, the 2-year smolt cycle is most common. Some smolt production facilities in Norway have been able to raise their winter water temperature by several degrees by pumping 6 to 8°C seawater from great depths in the Norwegian fjords.[353] The fry are thus exposed to a gradual increase in salinity from 6‰ in the fall to 25‰ the following spring. This has increased the proportion of 1-year smolts considerably.

3. Water Flows and Densities

Only small amounts of water are needed for hatching. In California-type hatching troughs with seven trays, each containing 1 to 1.5 l of salmon eggs, it is common to use approximately 10 l/min.[381] Hatching cylinders with 25 l of eggs, which are only used up to the eyed stage, require a flow of 7 to 8 l/min. These figures assume modest hatching temperatures; some increase in flow would be needed for the highest appropriate hatching temperatures.

If water under pressure is piped to hatching trays, air bubbles can accumulate under the perforated bottom plates of the trays and cause mortalities. To prevent this, the water is commonly aerated just before entering the trays.

Necessary water flows and densities during rearing vary a great deal depending on temperature. Cold water has more oxygen than warm water and lower rates of flow are thus needed as temperature declines. The necessary flow rate also varies with fish size. During startfeeding, when flows are small, production is usually limited by rearing area. In shallow tanks, which are most commonly used for startfeeding, it is usual to have 10,000 fry per m^2 of tank area. At 2 g average weight, densities of 3 kg/m^2 are common. That density is increased to 5 kg/m^2 at 10 g.[378]

As the fish grow they are frequently put into deeper tanks, where densities are more properly measured in terms of weight of fish per unit of water volume. Densities in those tanks vary from 3 kg/m^3 at 2 g to 15 kg/m^3 at 10 g and 25 kg/m^3 at 40 g.[378] The above figures apply to rearing temperatures over 10°C.

Various formulas have been developed for establishing water flows during smolt rearing. At 5°C, the water requirement per kilogram of salmon parr varies from 0.6 l/min at 1 g average weight to 0.4 l/min at 10 g and 0.3 l/min at 25 g.[382] At 10°C temperature the respective figures are 1.3 l/min for 1-g fish, 0.9 l/min for 10-g fish, and 0.5 l/min for the 25-g fish. Water flow requirements increase considerably at higher temperatures.

4. Mortalities

In the early days of smolt rearing it was assumed that one would lose approximately 50% of the number of eggs initially available through the hatching and rearing process to 2-year smolts. With better feeds, shorter rearing cycles, and better hygiene, it is more common to retain over 60% of the original crop. There is, however, great variation in total mortalities between years and from hatchery to hatchery depending on a variety of factors such as origin of stock, water quality, temperature, and egg size. As a rule, larger eggs give better survival during early feeding.

To be conservative, one can expect 15% mortality during stripping and hatching, an additional 20% during startfeeding, and some 5% during a year of feeding. That totals a mortality of 40%, though survivals up to 70% are not uncommon.

B. HATCHERY PRACTICES

Hatchery practices are fairly similar for most of the salmonid species, at least those with fairly long freshwater residency. All of the species under culture are fed dry or moist diets up to the smolt stage. The fish are graded and sorted at various times and great effort is put into proper hygiene. Basic hatchery practices used for Atlantic salmon in Europe are highlighted in this section.

1. Spawning

Wild Atlantic salmon in Northern Europe spawn during the period from September through November. The same spawning period basically holds for ranched salmon, but net-pen, cultured salmon are frequently spawned through December. Females are commonly spawned by manual extrusion of the eggs once the fish are ripe. Atlantic salmon can be reconditioned to spawn each year, and some brood fish can be very valuable when used repeatedly as a part of a selective breeding program.

Ranched salmon in Iceland used to be spawned by extrusion of their eggs, but when bacterial kidney disease (BKD) appeared in the stations it was decided to adopt the method used for the Pacific species and obtain the eggs by way of abdominal incision. The males are, however, spawned in a conventional manner. Both males and females are then slaughtered and checked for any incidence of BKD. Gametes obtained from infected brood fish are discarded.

Atlantic salmon eggs are variable in size, depending on the size of the female and the environmental conditions which prevailed during egg development. The volume occupied by the eggs varies from 4,000 to 12,000 eggs per liter, with an average around 7,000 eggs per liter. The quantity per female varies considerably, but is frequently 0.3 to 1 l in grilse females (2 to 3 kg), but 1 to 2 l in larger females (5 to 7 kg).

The most common method of fertilization is the dry method, where the eggs are fertilized in the absence of water. After being mixed thoroughly, the fertilized eggs are washed in water and allowed to water harden. Usually 1 to 2 hours are allowed for water hardening after which the eggs are placed in the incubation units, either hatching trays or hatching cylinders. In most cases the eggs are disinfected with an iodine solution before being placed in the incubation units.

2. Care of Eggs

During the water hardening process, the eggs are very sensitive to handling. After they are completely hardened, the eggs can be transported at low temperatures for considerable distances. After about 50 degree days the eggs become very sensitive again and by that time the initial picking of dead eggs should have been completed. Icelandic experience has shown that careful picking of eggs from trays can be accomplished throughout the hatching period, if the trays are not moved.

About halfway through the hatching period (250 degree days) the eggs reach the eyed stage and are shocked by pouring them from the hatching trays into a bucket of water. The eyed eggs are very hardy at this point. Frail and unfertilized eggs turn white after being shocked and can be removed. After the shock treatment the eggs can be shipped up to 48 h in special containers with ice on top to keep them cool.

It is very important to inspect the hatchery frequently and pick out dead eggs which will support the growth of fungus (*Saprolegnia* spp.) which can spread to otherwise healthy eggs. Malachite green is frequently used in hatcheries to combat fungus and a regular treatment with that chemical twice a week is necessary when eggs are hatched in large quantities in cylinders where egg picking is impractical. [Editor's Note: The use of malachite green is currently prohibited in conjunction with fish production in North America.] As previously mentioned, it is normal to lose 10 to 15% of the original egg quantity during the hatching process.

3. Feeding

Startfeeding, when the young fish are learning to take artificial feed, is the most critical part of the hatchery cycle. Some fry never learn to take feed and die. The fish are usually placed in startfeeding tanks when two thirds of the yolk sac are absorbed, at which point they start searching for food. They usually stay close to the bottom waiting for the food to drift by.[383] It is only after 2 weeks of successful feeding that healthy fry begin to occupy more of the water column. Mortalities typically begin to appear after 2 to 3 weeks of feeding and by the 6 week the mortality rate declines.[378] Normal mortalities during this initial period of loss are 10 to 20% but can occasionally be much higher depending on the egg quality and various other factors associated with husbandry.

Atlantic salmon are usually startfed in shallow square tanks, either 1 m^2 (1 m × 1 m) or 4 m^2 (2 m × 2 m). The smaller tanks are usually preferred if larger tanks are available once the fish reach 1 to 2 g. Smaller tanks make it easier for workers to inspect the fish and clean out dead fish, excess feed, and waste products. Hatchery managers usually try to keep good records of their mortalities at this point and throughout the rearing process as this is the most practical way to predict final smolt production.

Some commercial operators have managed to startfeed their fish in large production units, keeping the initial water level low and increasing it as the fish grow. The basic principles, however, are the same. Water depth during startfeeding is usually 15 to 25 cm. Depth can be adjusted to that which is optimum for the tank once all the fish have established pelagic feeding behavior.

Atlantic salmon are normally startfed with commercially available dry diets which have usually proven better than beef liver and other wet feeds which were used in the past. The dry feeds are often high in fat and require specially designed feeders. In North America, moist feeds and beef liver mixed with dry diets have given better results than dry diets alone.[380] The differences between the European and North American experiences may be due to the fact that the quality of starter feed is variable between feed companies while its composition is more critical than the feeds used at later stages of production.

Startfeeding with dry feeds is usually done with automatic feeders which distribute feed 5 to 10 times per hour. Additional handfeeding is sometimes practiced. It is very important during the startfeeding period to feed in moderate excess of the amount that the fish consume. Feed conversion rates are usually in excess of 3.0 (kg of dry feed required to produce each kg of gain) during this period compared to 1.0 to 2.0 for smolts and later stages. Although starter feeds are relatively expensive, one should avoid underfeeding as it can have serious consequences.

Once the fry reach 1 to 2 grams they are frequently transferred to larger tanks (4 to 16 m^2), which may be either shallow square ones, originally of Swedish design, or fairly deep square or circular tanks. The deeper tanks, which allow water depth of approximately 1 m, are becoming more common in modern stations as they provide greater rearing volume per rearing area. This is especially important in colder climates such as those of Iceland and northern Norway, where these tanks are usually inside a building. In southern Norway and Scotland the tanks are usually kept outdoors, and often are fitted with a semitransparent lid. The fish may be maintained in those tanks until smolt size, although some stations have larger outdoor tanks with ambient water for production of smolts.

Dry feeds and automatic feeders are almost exclusively used up to the smolt stage. The feeders are controlled by electronic timers and are set for various feeding levels and frequencies. It is considered best to offer small amounts of feed at frequent intervals. Feed is generally offered both day and night. Some managers, however, set their feeders to deliver most of the feed at dusk and dawn when the fish seem to be most actively feeding. Convenient feeding tables, which specify the correct particle size and amount to be fed for a certain temperature and a certain fish size, have been designed by the feed companies. The quantity of feed offered per day is usually specified as a percentage of the total weight of fish in the tank. Although feed tables are useful as an aid in planning the daily routine, nothing can replace an observant hatchery manager who knows when the fish are properly fed and in good health.

4. Grading

Salmon fry are usually graded several times during the rearing period. Size variations within a tank are soon apparent and experience has shown that the larger fry suppress the growth of the smaller fish, though compensatory growth will occur if the small fish are placed into another tank with fish of similar size.

Grading is usually accomplished with a manual fish grader which is basically composed of a rectangular box with variable size fiberglass bar screens in the bottom. The screens allow the smaller fish to escape while the larger ones are retained. Large commercial operations often use commercially available fish graders.

Atlantic salmon are commonly sorted when they reach 6 to 7 cm in length. The need for grading is often defined as occurring when some of the parr in a tank are outside of 10% variation from the mean length.[384] Under normal circumstances it is satisfactory to grade Atlantic salmon every 3 months up to the smolt stage.

5. Hygiene

Successful smolt production depends to a large extent on cleanliness and proper disease control. Dead fish must continually be removed and excess feed particles and excrement scrubbed from the tanks. Circular and semicircular rearing tanks with a central drain, which dominate Atlantic salmon production, are relatively easy to keep clean and in some cases are largely self-cleaning. Mechanical scrubbing of tanks cannot be avoided, but should be kept to a minimum so as to not stress the fish unduly.

In order to prevent disease and parasite transmission from tank to tank, each rearing unit is usually equipped with its own brush and a small perforated spade to pick dead fish. These, along with dipnets, should not be transferred between tanks without first receiving thorough disinfection.

Various treatments have been developed for disease or parasite epizootics in rearing tanks. In the case of parasites and some bacteria, treatment baths with formalin or malachite green are commonly used. The fish can be fed antibiotics for the treatment of certain bacterial infections. Vaccinations have been developed which are effective against certain bacteria and viruses.

C. REARING PROCESSES

There is a great variety among the rearing procedures used in the production of Atlantic salmon smolts and adults, depending on latitude and local conditions. One particularly important variable is the water source used and the availability of hot or temperate water to enhance the rearing process. In countries such as Sweden, which rely on ambient water temperatures, smolts are primarily produced in 2 years, while in more temperate climates such as Ireland and France, most of the parr reach smolt size in 1 year. In other countries, such as Iceland, which depend on the heating of rearing water with geothermal energy or industrial effluent, there are possibilities to rear parr at constant temperatures. Under those conditions the salmon reach smolt size in early winter and can be reared to a size of 400 to 600 g the following spring. These fish are probably not suited for ranching and the constant temperature rearing may have peculiar effects on the timing and expression of smoltification. The latter type of rearing cycle may, however, have a major potential for enhanced rearing of salmon which remain captive throughout their life cycle.[379]

This section describes and compares some of the rearing processes which are currently in use and shows how smolts for ranching and captive rearing can be produced.

1. Two-Year Smolts

When smolt rearing started in Sweden in the 1960s, production was based solely on a 2-year cycle. The Swedes basically used river water at ambient temperature for the entire rearing period. This meant that they had to adapt hatching and startfeeding activities to local temperature conditions. The parr had fundamentally 2 summers to grow, with only minimal growth occurring during the winter. The parr were kept indoors until the end of their second summer, at which point they were put in large outdoor ponds for overwintering. Low winter temperatures and natural photoperiod for almost a year ensured proper smoltification, and the resulting smolts, which often were 50 grams in weight, produced good returns to the Baltic Sea fisheries. Similar smolts are still used for delayed releases in the Baltic area and return rates have continued to increase.[385] In the 1970s and early 1980s considerable numbers of smolts were exported from Sweden to various salmon farms in Norway. Many countries in Europe and North America followed Sweden in the production of 2-year smolts, although many have since converted to 1-year smolt production.

2. One-Year Smolts

Several countries with rearing facilities in temperate latitudes can easily produce 1-year smolts, even with ambient river water. Such smolts have been produced since the early 1970s in Ireland.[356] Natural rearing conditions in Iceland do not favor 1-year smolt production. However, when geothermal energy was harnessed for salmon production in the late 1960s, the opportunity existed to concentrate entirely on a 1-year smolt cycle. Initial trials

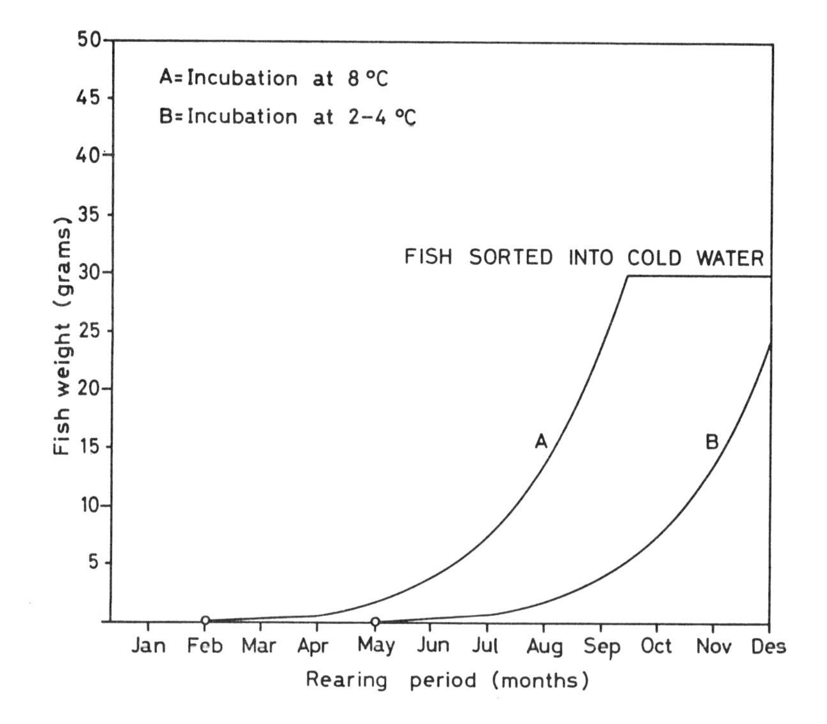

FIGURE 28. Growth rate and rearing schedule of accelerated smolts (A) and smolts hatched in cold water (B) in Icelandic smolt farms. Growth rates correspond to 12°C rearing temperatures.[379]

indicated that 1-year smolts were inferior in rearing and ranching to their 2-year counterparts.[386] These problems turned out to be related to the use of improper photoperiods and improper handling. By 1973, 1-year smolts were performing comparably to 2-year smolts in ranching trials.[387] Icelandic 1-year smolts are 30 to 40 g in size, considerably smaller than 2-year smolts used previously, and smaller than most smolts used in aquaculture elsewhere.

Figure 28 compares the 2-year cycle with the 1-year cycle used at Kollafjördur Experimental Fish Farm in Iceland, demonstrating that the conversion to 1-year smolts was primarily accomplished by advancing startfeeding by 3 months. This made it possible to rear over 80% of the production to smolt size before January 1 of the year of release, allowing ample time for proper photoperiod treatment before release.

Until 1970, Norwegian smolt production was primarily 1- or even 3-year smolts. The proportion of 1-year smolts has been increasing greatly and today a large share of the smolts produced in Norway are 1-year olds.[388] This is primarily due to enhancement of the rearing cycle during hatching and rearing by heating the rearing water, either through the use of industrial waste heat, heat pumps, or by pumping temperate seawater from great depths during the winter. Similarly, the proportion of 1-year smolts increased in Scotland from 50% in 1982 to 85% in 1985.[364]

The actual routine of producing 1-year smolts varies considerably between countries, but the underlying principles are the same. Figure 29 shows a suitable rearing process for 1-year smolts developed at the Kollafjördur Experimental Fish Farm in Iceland.[389]

Eggs are typically taken in October and hatched in 8 to 9°C water. This means that the fry are startfed in January and February and are intensively fed up to a size of 20 to 30 g. During this period the parr are fed day and night in subdued light to get maximum growth. The most rapidly growing parr reach smolt size in early August, but the slower growing

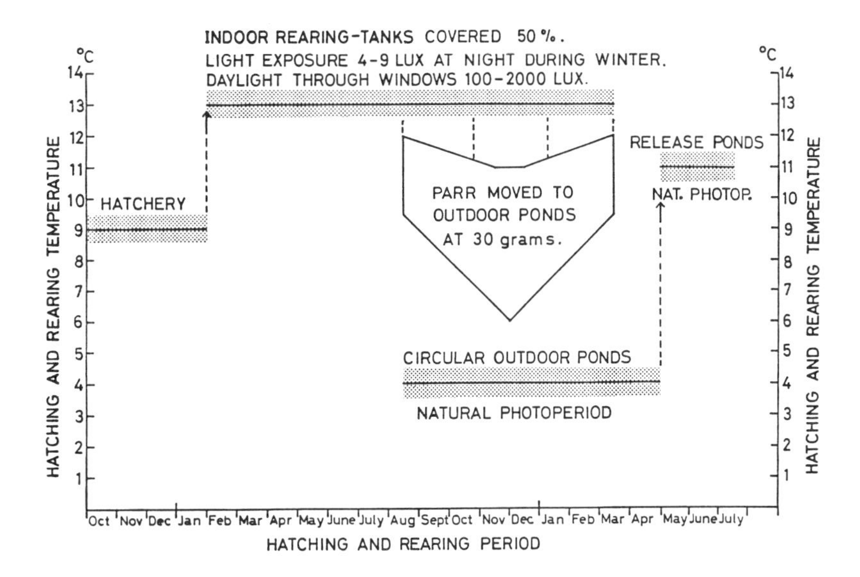

FIGURE 29. Hatching and rearing programs used for the production of 1-year Atlantic salmon smolts at Kollafjördur Experimental Fish Farm in Iceland, emphasizing appropriate temperature and photoperiod treatments during the rearing period.[389]

fish do not reach that stage until early March. Early smolts are moved into outdoor ponds which receive 4°C well water for overwintering in natural photoperiod conditions. In the spring, just before release, the smolts are normally 30 to 40 grams.

The rearing cycle of 1-year smolts in other countries is a variation of the program described above. Many smolt stations advance the hatching and early feeding periods to match the rise of ambient temperature in the spring. In Norway, the rearing temperatures during the first winter are frequently elevated through the use of temperate seawater obtained from depth in the Norwegian fjords. This practice was, however, banned by Norwegian authorities in 1990 to reduce the risk of disease. In other cases warm industrial effluents have been used to advance the rearing process.

3. Zero-Smolts

Enhancement of the rearing process opens various avenues for shorter smolt rearing cycles and with enhancement at later stages for an even greater shortening of the total rearing cycle. One alternative, which has been exploited with coho salmon, is the production of 6-month smolts.[390,391] The smolts, which are often called zero-smolts as they are produced without completing a winter in freshwater, have been successfully used in an aquaculture venture in Oregon.[392]

Zero-smolts of Atlantic salmon have been experimentally produced in Iceland in small quantities. Unfortunately these smolts have usually only been ready for release in July or early August, after the normal migration period for smolts. Although these smolts have produced poor returns in ranching trials, they have performed well in salinity tests and should be suitable for captive salmon rearing. The only way to enhance this process further is to speed up the maturation of brood fish. Some experimental groups exposed to artificial photoperiod control have provided eggs in mid-July. Eggs from those females should provide an additional boost to zero-age smolt production in Iceland.

4. Enhanced Rearing Cycle

If parr destined to be zero-smolts are kept for an additional year at optimum temperatures

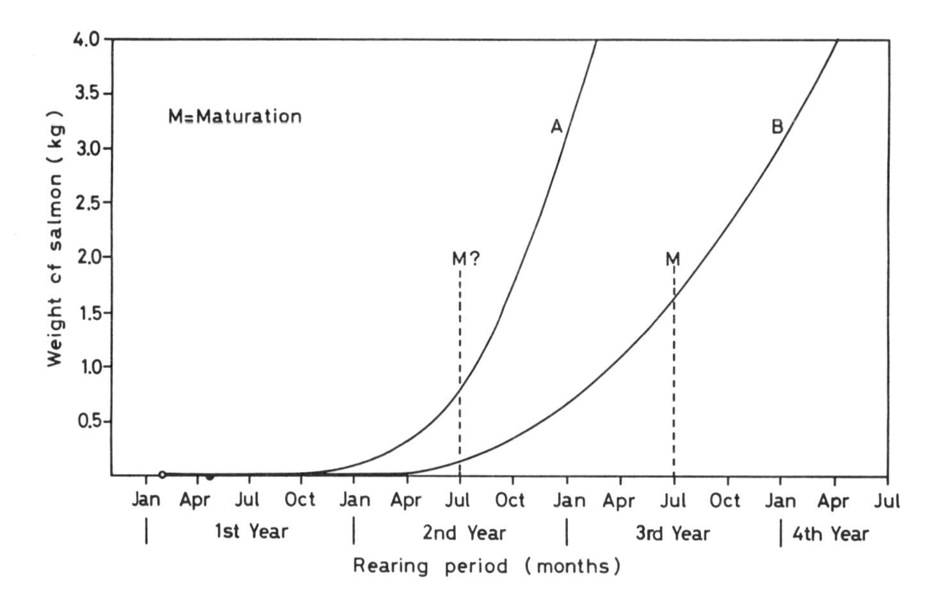

FIGURE 30. Comparison of projected growth rates for accelerated rearing of Atlantic salmon to market size at 12°C in land-based tanks in Iceland (A) and conventional rearing in sea-pens at ambient temperatures in Norway (B). M indicates the likely time of sexual maturation. Early maturation has been a major obstacle in accelerated rearing programs in Iceland.[379]

for growing, they can reach a size of 400 to 600 g as 1-year smolts.[379] These fish are considered too large for release in ranching operations, but can be reared to market size, either in pens or land-based tanks. The cycle has sometimes been interrupted in experiments by early maturity in 1 to 1.5-kg salmon during the first summer. If maturity could be prevented through selection, sterilization, or other techniques and the fish kept at optimum temperature throughout the cycle it might be possible to speed up the rearing cycle by a full year compared to the conventional rearing cycle in pens stocked with 1-year-old smolts (Figure 30).

It seems likely that this increase of a year in turnover rate in conjunction with land-based fish farms (e.g., in Iceland) is the only justification for the additional investment costs in tanks, pumps, and heating systems. Net-pen farmers could also possibly plant 400 to 500 g post-smolts in the spring and slaughter in early winter in order to avoid overwintering the salmon in the pens in areas where temperatures are critically low and weather conditions harsh.

V. SEAWATER REARING

The term "seawater rearing" is defined as the process of producing Atlantic salmon foodfish in pens, land-based units, or other enclosures. Most of this production uses full strength seawater, but freshwater or brackish water production of salmon has been successfully attempted under a variety of salinity regimes. One of the most successful net-pen operations in Iceland has utilized a lagoon with salinities as low as 10‰.

Physical and biological parameters of salmon farming are described in the first part of this section. Rearing practices, including feeding practices and disease treatments, are considered thereafter.

A. PHYSICAL AND BIOLOGICAL PARAMETERS

1. Natural Conditions

The local conditions in each country logically shape the rearing methods used in that country. It is probably not a coincidence that Norway was a pioneer in Atlantic salmon culture nor that it has had the most rapid development. The shoreline of Norway is especially well suited for the rearing of salmon in inexpensive net-pens.[394] In addition to a great number of islands and skerries sheltering the coast, there are moderate tidal amplitudes in the best locations and the Norwegian fjords are very deep, even along the coastline, which allows anchoring of pens close to shore. Winter sea temperatures rarely fall below 4°C and summer temperatures are moderate. Additionally, the marine fisheries are well developed, providing large quantities of scrap fish, and the political environment favors new industries in remote communities along the coast.

In Scotland, which was not initially far behind Norway in fish farming development, salmon farming has never gained comparable momentum. As the climate in Scotland is more harsh than in Norway (more intense storms), pens were more expensive and less practical. Some farmers consequently turned to land-based rearing. Similarly, Ireland and the Faroe Islands lack extensive sheltered areas and had to rely on advanced high-seas net-pen technology for their salmon farming. Iceland, in addition to a harsh climate, was faced with the coldest sea temperatures and many fish farmers either established land-based units or had to rely on very sturdy oceanic pens.

It might be argued that Iceland has the best natural conditions for land-based salmon rearing, as some areas of the country have ample geothermal resources which can be used to enhance the rearing process. In some areas, seawater of suitable temperatures can be pumped directly from boreholes, but in others ambient seawater would have to be heated with steam or high temperature geothermal water. This assumes that elevated temperatures would be advantageous throughout the rearing process.

2. Rearing Temperatures

It is well known that young salmon in freshwater grow best at temperatures from 10 to 15°C. The most suitable temperature range for the growth of post-smolts in seawater is less well defined. Norwegian information indicates that the upper limit is around 20°C and growth is very slow below 4°C.[369] Experience has, however, shown that Atlantic salmon culture is most successful in areas where temperatures do not exceed 14 to 15°C.[394]

Subzero temperatures are lethal for salmon, if they are below −0.5°C, which is more than 1°C above the freezing point of normal seawater. Icelandic experience has shown that Atlantic salmon can tolerate −1.5°C seawater for short periods if not disturbed or otherwise stressed.[395] Mortalities are common in supercooled seawater in which surface ice has not developed. Occasional winterkills have occurred in net-pens in Northern Europe, and this danger actually precludes net-pen rearing in some areas of southern Norway, where cold seawater from the Baltic penetrates into the Atlantic in the wintertime. Similar conditions exist in eastern Scotland, eastern Canada, some parts of Iceland, and northern Norway. It should be stressed that the best temperatures for salmon farming exist in areas which are directly in the path of the Gulf Stream, which warms large areas of northern Europe.

Favorable temperatures also exist in the Pacific Ocean along the coasts of the state of Washington in the U.S. and the province of British Columbia in Canada, where net-pen rearing of Atlantic salmon and Pacific salmon is gradually increasing.

3. Densities

The allowable densities of salmon in pens are variable, depending on local conditions, primarily tidal exchange. In land-based units, where the water is constantly pumped, the densities are best based on flow rates. It is conservative to estimate that the necessary flows at 10°C would vary from 0.5 l/min/kg for smolt-size fish down to about 0.2 l/min/kg fish for salmon of several kilograms. Densities up to 35 to 40 kg/m^3 have been reported in Icelandic land-based operations,[396] but it seems that those densities have rarely been obtained during practical operation. In land-based operations oxygen injection is possibly a viable alternative to increased pumping activity and densities can probably be considerably higher than in net-pens, especially if rearing temperatures can be rigorously controlled and oxygen concentrations optimized.

In net-pen rearing it is common to plant smolts at densities of about 1 kg/m^3, which corresponds to 20,000 smolts in a typical Norwegian net-pen of 500 m^3.[369] After a few months, the fish are redistributed into three pens. Average densities have, in general, been increasing as growers gain experience and knowledge. Some very high densities (for example, 50 kg/m^3)have been reported, but more realistic densities are 15 to 20 kg/m^3.[369] In general, one can expect more disease and stress-related problems as densities increase. Considering the relatively small investment in the pens and assuming ample natural rearing space in a fjord, great increases in density could in the long run be of questionable value.

4. Mortalities

In net-pen culture it is difficult to separate natural mortalities from other sources such as escapes from the pen or predation by birds or other animals. A fish farmer can lose an entire crop in 24 hours through storm or seal damage to the netting. Mink, otter, and various species of birds can remove significant numbers of fish in a short time if their activity is unchecked. Fish, on the other hand, that die from natural causes quickly disintegrate in seawater and are lost from the pens. For practical purposes one must thus work with approximate values of survival which take into consideration various factors, excluding total loss. In Norway, it is assumed that 10 to 25% of the smolts that are stocked will be lost before harvest.[369] The rate of loss is greatest during the first year (15 to 20%), but 3 to 5% of the standing crop are lost in the second year of rearing. In order to stay within those ranges the fish farmer must regularly check the nets and repair any damage as it occurs.

Diseases, which are discussed in more detail below, take their toll in salmon farming both in pens and in land-based units. Fish which are kept at high densities during their rearing cycle are highly susceptible to disease, especially if environmental conditions in the rearing unit become suboptimal. Diseases arise from parasites, bacteria, and viruses as well as from faulty nutrition, which is less common since the advent of balanced fish diets.

In addition to these factors it is not uncommon to have unexplained losses which may primarily arise from poor initial fish counts or bookkeeping errors. Smolts are difficult to handle and numbers are frequently estimated from biomass. This method can be fairly inexact if there is a great variation in the size of smolts.

5. Growth and Maturity

Growth of cultured salmon varies widely depending on local temperature conditions and whether the rearing takes place under natural or enhanced production cycles. The bulk of Atlantic salmon rearing takes place in net-pens under ambient temperature conditions. The following information was obtained from the net-pen farming industry in Norway.

Thirty-gram smolts leaving a river in the spring can increase in weight by one hundredfold (reaching as much as 3 kg) before they return to spawn after 1 year at sea. The growth of salmon in captivity is generally slower, even though smolts are usually over 40 g at stocking, considerably larger than their wild counterparts. An average size of 2 kg after 1 year of rearing is considered satisfactory. Growth during the second year is relatively greater and average weight after 2 years of rearing is frequently 6 to 8 kg.[369] Since salmon of 2 to 3 kg are considered too small for marketing, the production cycle in most operations is 1.5 to 2 years. This cycle can be successfully maintained if the salmon do not mature in the first summer of their life. When salmon mature their growth rate slows and they become discolored and unfit for market in late summer.

Early maturity has frequently been a problem in salmon culture. In some cases it has been the factor that decided between success and failure of a rearing operation. The process is related both to genetics and environmental conditions. Many wild stocks have an inbred tendency to mature early while fish farmers usually try to find brood stocks with late maturation. In the early 1970s when selection programs to breed better salmon for rearing were started in Norway, late maturity was one of the main initial objectives.[397]

The role played by the environment is also significant. Early maturity is much more common in areas with relatively high winter seawater temperatures, such as southern Norway, Ireland, and Scotland, but less common in northern Norway and northern Iceland. In some cases there have been pronounced increases in early maturity following mild winter temperatures, even in selectively bred Norwegian stocks.[398]

B. SPECIFICATION AND MAINTENANCE OF NETTING

In Norway the netting used to enclose the fish is made from nylon fish netting with a stretch mesh size of 10 to 12 mm for smolts and up to 15 mm for 100- to 500-g fish. For larger salmon, the mesh size may range from 20 to 26 mm.[369]

Net-pens are not very costly compared to other types of rearing facilities, but they require a considerable amount of maintenance. The nylon netting tends to become fouled with marine growth, including seaweeds, bivalves, and various other types of organisms. If this is allowed to go unchecked the circulation of seawater through the net-pen may become blocked and can not sustain high densities of fish. Fouled nets can be cleaned with high pressure water jets, but it is more common to change the netting regularly, e.g. twice per year.[353] These changes usually take place in the spring and late summer as fouling is much greater during the warm summer months.

In order to discourage fouling, the net bags are normally impregnated with copper compounds during each cleaning operation. Nets available commercially are usually treated. Smolts, however, are sensitive to these compounds and initial planting is usually into nonimpregnated netting.

C. THE REARING ROUTINE

In salmon culture it is important to have a work plan for the various seasons in order to utilize personnel in an efficient manner. Some of the activity such as the planting of smolts in the spring and spawning of salmon in the fall is seasonal, but other activities may occur daily or at irregular intervals. The rearing routine in Norwegian net-pen culture has been extensively described by various authors.[353,369,399]

Normally the smolts are put into pens in the spring or early summer. The netting must thus be mended and prepared for the new season. The stocking of smolts represents a very critical time in the rearing process, as the smolts may have difficulty in adapting to seawater, especially after a stressful period of transport. One has to be especially observant regarding disease outbreaks at this time and any change in fish appetite or increased incidence of mortality should be carefully monitored.

If the smolts start feeding normally, the quantity fed increases rapidly as the sea temperature rises and the fish grow. The main job during the summer period is to feed the fish, either manually or with automatic feeders which are commonly used with dry feeds. The net bags have to be kept clean and net areas frequently changed during late summer. Farm personnel must count and remove dead salmon from the pens at least once a week in order to keep track of the number of salmon in each net-pen.

The temperature of seawater drops slowly in the fall and the salmon can feed well into November or even into December, depending upon local climatic conditions. As temperature drops in the fall, it is very common for farmers to grade the salmon that were planted the previous spring. It is also very important for the culturists to check the condition of the pens and their moorings before the advent of the winter storms, as most disasters due to heavy winds, high seas, or ice occur during that period.

As winter progresses and the temperature of seawater drops, the appetite of the salmon is sharply reduced. At temperatures around 1 to 3°C, fish on dry feeds usually stop feeding sooner than those on moist diets due to the high salt concentration of the dry feed relative to water.[369] Salmon have difficulty disposing of excess salt at low temperatures and some mortalities may occur. Wet feeds, which have a ratio of salt to water similar to that of the growing salmon, do not create this problem and are preferred at low temperatures.[400] In Icelandic feeding experiments, some feeding was observed even at subzero temperatures in seawater cages.[395]

Salmon have traditionally been slaughtered in Norway in the spring, but there is an increasing tendency to distribute the period of harvest to other parts of the year, primarily late fall and early winter.

VI. MAJOR PROBLEMS ASSOCIATED WITH REARING

The problems associated with salmon rearing can primarily be devided into three groups. First, there are those directly related to husbandry either through stress, diseases, food quality, or combination of those factors. Those problems can be thought of as being of internal origin. The second group relates to external factors that can threaten an otherwise healthy stock of fish in captivity as a result of exposure to pollutants, severe weather, or other environmental disturbances. A third group discussed here deals with environmental problems posed by the aquaculture industry on the culture environment. Examples include self-pollution and genetic mixing of cultured stocks with wild stocks.

A. INTERNAL PROBLEMS

Although discussed here separately, the internal problems in salmon culture are frequently associated with some changes in the external factors that impinge upon the rearing environment. Diseases, for example, are often associated with abnormally high temperatures or some deterioration of the rearing environment. These effects can, in some cases, be counteracted through proper husbandry.

The physiological state of the fish is of major importance, considering that salmonids have to make a transition from freshwater to seawater and undergo a major physiological change during that transition. The major internal problems discussed here are those related to physiology, fish diseases, and general husbandry techniques.

1. Physiology

Physiological problems are common in Atlantic salmon culture, both in rearing and ranching programs. Numerous smolt rearing stations have produced juveniles which either did not smolt properly or smolted at the wrong time of year. Such smolts will not migrate to sea in ranching operations and do not tolerate the transfer from freshwater into net-pens. Their subsequent growth is retarded and mortalities may be abnormally high.[401] Proper smolting and smoltification timing is thus of utmost importance both in salmon farming and ranching.[402]

Another example of a physiological problem is the early maturation of some salmon stocks and the precocious maturity of salmon males.[138,403] These problems can be so severe in Atlantic salmon culture that the economic success of the operation may be jeopardized. This is demonstrated by the emphasis on selection for late maturity in Norwegian breeding programs after it became clear that the Norwegian farming cycle was only economical if the salmon were grown in the pens for 1.5 to 2 years from the time of smolt planting.

The problem of early maturation is also being dealt with by making triploid sterile salmon either by exposing newly fertilized eggs to a heat shock or high pressure.[404] Experimental production of triploid salmon is ongoing in many countries.

2. Fish Diseases

Diseases can be a major problem in a salmon farming operation. Numerous diseases have been found in cultured salmon and new diseases are frequently being described. Most of those diseases are probably found in natural populations in the respective areas where culture is practiced and are gradually being transmitted into the rearing environment, but in some cases very serious diseases have been transported between countries or continents.

The frequency of disease in salmon rearing as in any other type of intensive culture is primarily related to the high densities under which the salmon are being produced. Other factors of importance are the relative presence of fish pathogens in the water supply or rearing water as well as general sanitary conditions in the rearing operation.

Fish diseases can primarily be broken into four groups:[405]

1. Infectious diseases, mostly bacterial or viral
2. Parasitic infections
3. Nutritional diseases
4. Environmental diseases

a. Infectious Diseases

Infectious diseases are attributable to bacteria or viruses. Bacterial infections can, in many cases, be treated with antibiotics. Viruses, on the other hand, utilize their hosts cells to reproduce and can not be treated by chemotherapy.[406] Several bacterial and viral diseases have caused problems in Atlantic salmon culture. Some of the more common diseases are discussed below.

i. Furunculosis

Furunculosis, which is considered the scourge of salmon culture, is caused by the bacterium *Aeromonas salmonicida*. External signs of the disease are red boils at various locations on the body. Similar boils and hemorrhages may also be found internally. It is found in salmonid culture operations over much of Europe and in North America, where it is often the cause of significant losses.

Furunculosis recently became a problem to Norwegian salmon culturists after infected smolts were imported from Scotland.[407] The disease had previously been reported in Norwegian rainbow trout culture and a few smolt stations.

Several antibiotics have been used to treat and control furunculosis with varying degrees of success. Efforts should be made to eliminate the disease wherever it occurs and to reduce its geographical distribution.[406] In spite of 30 years of research, no vaccines have been developed against furunculosis.

A subspecies of furunculosis, *Aeromonas salmonicida achromogenes,* has been found in salmon culture in various countries, including Iceland, where the common strain has not been found. *A. s. achromogenes* has appeared to be somewhat less virulent to cultured salmon than the related strain.

ii. Vibriosis

Vibrio anguillarum is a saltwater bacterium which is of great economic importance in marine net-pen culture, where it may cause great losses. The bacterium causes external and internal hemorrhaging similar to that of furunculosis. *V. aguillarum* is widely distributed in the marine environment throughout the world, and over 50 species of fish are known to be susceptible, ranging from salmon to turbot.

In Norway, outbreaks of vibrio are especially common just after the salmon have been exposed to high temperatures or other stresses such as transport, grading, or transfer from freshwater to saltwater.[405]

As the bacterium is not an obligatory parasite, disease outbreaks are less likely if the rearing conditions are kept as optimal as possible for the species in question. Many antibiotics have been found useful to treat vibriosis in salmon and vaccinations which provide protection from the disease have been developed.

Since 1980 a new strain of vibriosis, *Vibrio salmonicida,*[408] has caused considerable losses in Norwegian net-pen rearing operations. It has been named "cold-water vibriosis", as it appears to break out during periods of very low seawater temperatures.[405] This disease has increased in importance compared to conventional vibriosis in European salmon culture in recent years, although considerable success in controlling the problem has been achieved through vaccination.[408]

iii. Bacterial Kidney Disease

BKD is caused by the bacterium *Renibacterium salmoninarum,* which is a obligate parasite and thought not to survive well outside its host.[406] The bacterium is fairly well adapted to its host and very often occurs in a chronic or latent form.

BKD has widespread distribution in North America and Europe, where it was first reported in wild salmon from the Dee River in Scotland in the 1930s. External signs of the disease are few, apart from listlessness and lethargy. Internally, infected fish often show gray-white abscesses in the kidneys or other visceral organs.[406] Mortalities are often slow to develop, even in highly infected populations. Cumulative long-term mortalities, however, may be great.

BKD has been of growing importance in European salmon culture in recent years, particularly in conjunction with ocean ranching operations in Iceland where the disease is frequently observed in latent form in Atlantic salmon brood stock. This is of special concern as the bacterium is considered to be transmitted from one generation to the next through the eggs. The current strategy involves examination of all wild and ranched salmon brood fish for BKD. Eggs obtained from infected individuals are destroyed. This technique, along with thorough disinfection, has allowed clean-up of heavily infected smolt stations.

BKD is one of the most difficult bacterial diseases to control with antibiotics as the disease is chronic in nature and the bacterium may be intracellular. The bacterium is, furthermore, not very sensitive to the most commonly available antibiotics.[406] Prevention is thus the best method of controlling BKD and movement of infected fish should not be allowed into noninfected geographical areas.

iv. Other Bacterial Diseases

Other bacterial diseases of importance are gill disease and bacterial fin rot. Gill disease is frequently found in fry and juveniles at smolt production facilities. These diseases are caused by so-called flexibacteria, which usually attack the gills after they have been irritated by feed particles, silt, or chemical substances.[409] Bacterial fin rot is usually a secondary infection which occurs after fins have been damaged mechanically or have been infected by a bacterium.

v. Viral Diseases

There are three major viral diseases associated with salmonid aquaculture in Europe: VHS (viral hemorrhagic septicemia), IPN (infectious pancreatic necrosis) and IHN (Infectious Hematopoietic Necrosis). Two of those, VHS and IHN, are primarily associated with rainbow trout culture and have not caused significant mortalities in Atlantic salmon culture. IHN was first found in 1987 in Europe, a continent to which it was probably imported with eyed rainbow trout eggs from North America.[410]

IPN is primarily of concern in Atlantic salmon culture during the early phase of juvenile rearing. Mortalities associated with IPN are erratic. Infected smolt stations may have high mortalities in 1 year and very small losses in other years.[409] A new viral disease, SAS, also called infectious salmon anemia, has recently been described from Norwegian salmon farms.[412] There are no available chemotherapeutic methods available for the treatment of viral diseases, though research to develop effective vaccines is underway.

b. Parasitic Infections

Many of the parasitic epizootics which impact Atlantic salmon are restricted to the freshwater rearing phase, with the exception of sea lice. The most common diseases are caused by protozoans, various types of flukes, and copepods.

i. Protozoans

Costiasis is produced by the protozoan *Costia necatrix,* which can cause severe problems during the juvenile rearing phase of Atlantic salmon.[409] This ectoparasite gives bluish-white coloration to the skin of salmon fry. Similar problems can be caused by the external parasite *Trichodina* spp. These parasites can be effectively controlled by bathing the fish in a dilute solution of formalin. The protozoan *Myxosoma cerebralis* causes the infamous ''whirling disease'', which has long been a problem in European rainbow trout culture. The disease is primarily a problem in earthen ponds as the parasite needs to complete its life cycle in the bottom mud. Its importance in modern salmon culture is thus diminishing.

ii. Salmon Flukes

The most common fluke in Atlantic salmon culture is *Gyrodactylus salaris*. It is primarily a problem in juvenile rearing. Once the parasite has established itself within a hatchery system, it is very difficult to eliminate, but infections can be controlled by bathing the fish in a formalin solution or saltwater. This parasite, probably introduced into Norway from the Baltic, is infamous for the disastrous effects it has had on wild Norwegian salmon populations after spreading from rearing stations into salmon rivers. Over 30 salmon rivers have been seriously affected, and some wild populations have been nearly eliminated.[411]

iii. Copepods

The salmon louse (*Lepeoptheirus salmonis*) is the primary parasite of importance in seawater culture of Atlantic salmon.[405] The louse, which is frequently observed on Atlantic salmon soon after freshwater entry, is common on all salmon while they are in the sea. In salmon culture, where the salmon are densely crowded, the number of parasites present on individual fish can reach levels never observed in nature and may lead to damage and mortalities. The greatest problems are observed in net-pen culture where water exchange is poor.[409]

The salmon louse does not tolerate freshwater, and infections can be reduced by towing salmon pens into low salinity regions within estuaries. Methods have also been developed to bathe salmon in a solution of Neuvon, a phosphorus-based insecticide.[409] Other copepods such as species in the genus *Caligus* have also caused problems in salmon culture.

c. Nutritional Diseases

Nutritional diseases are of lesser importance to salmon culturists as the diets in current use are designed to meet the known requirements of the fish. Historically, salmon nutritional deficiencies appeared as a result of insufficient levels of vitamins and dietary lipid imbalance. The latter often caused fatty liver degeneration. Most presently used commercial dry feeds are of high enough quality to prevent any such disorders. These diseases would thus be most likely to occur when wet feeds and vitamin mixes are being used or after the feed has been improperly stored.

d. Environmental Diseases

Atlantic salmon need to have water within a certain temperature range, of the proper pH, high in dissolved oxygen, low in ammonia, and meeting a variety of other water quality limitations. It is commonly accepted that Atlantic salmon should not be exposed to temperatures in excess of 20°C for any length of time. Salmon have relatively high oxygen requirements, and the oxygen carrying capacity of water is reduced as temperature increases. At 20°C, for example, the oxygen level at saturation in fresh and seawater is well below 10 mg/l.[413] Similarly, the desirable pH of freshwater for salmon culture lies between 6.0 and 7.5.[405] At lower or higher values one can expect to experience disease problems in smolt-rearing stations. As many stations use run-off water from streams or lakes, there can be changes in water quality with time as a result of acid rain, for example, which is a growing problem in southern Scandinavia as well as North America.

Heating of rearing water in closed systems, without subsequent aeration, can lead to supersaturation. This can lead to the formation of nitrogen bubbles in the tissues of the fish with resulting damage and even death. The most common symptom is exophthalmia or "popeye", where salmon fry may lose one or both eyes. This disease can be prevented by proper aeration of the rearing water.[405]

It is also known that heavy metals such as copper and zinc, ammonia, and various other substances are highly toxic to salmon and can lead to mass mortalities within a short time. Heavy metals should be excluded from the water source as well as from any plumbing used. Ammonia is a metabolite that can be controlled by sufficient water exchange in open systems or biofiltration in closed water systems.

The rearing environment can be quite stressful for salmon as a result of one or more of the above factors. Insufficient feeding levels can lead to aggression and fin-nipping as well as attacks on the eyes. Fin deterioration is probably the most common disease in Atlantic salmon rearing facilities. Good husbandry and proper treatment of unhealthy fish are of prime importance in keeping the incidence of disease under control.

B. EXTERNAL PROBLEMS

It is quite clear that aquaculture enterprises are highly vulnerable to various changes that may take place in the environment, many of which are beyond human control. Acid rain can have major impacts on smolt facilities. Algal blooms, pollution, and harsh weather can destroy fish in net-pens, whereas pump and electrical failures have caused considerable problems in land-based operations.

1. Pollution

Pollution is here defined as any change that takes place as a result of human activity. The most important aspects that relate to salmon culture are acid rain, algal blooms, and oil spills.

Acid rain was previously mentioned as a disease factor in juvenile salmon production. It is primarily of concern in freshwater as seawater has a much greater buffering capacity. Acid rain is of considerable concern in the production of salmon smolts for Norwegian net-pens. The phenomenon is being intensively studied on both sides of the Atlantic and various mitigation measures have been attempted to diminish the impact of the problem, not only on fish culture operations but on natural fish populations.

Eutrophication of the marine environment seems to be a major threat to marine net-pen culture in some areas. This problem has been best demonstrated by the unusual blooming of several species of marine algae in the Kattegat/Skagerak region on the border between the Baltic and the North Sea.[414] Blooms have led to several instances of mass mortality involving various fish species in the area and salmon in Norwegian net-pens. Salmon have been killed as far north as Hordaland. The first reported incidences, which were from 1966, 1976, 1981, and 1982, involved the dinoflagellate *Gyrodinium aureolum,* which causes red tide.[414]

A more serious outbreak in 1988 involved the poisonous alga, *Chrysochromulina polylepsis.*[415] It was feared that the massive bloom, which travelled far north along the Norwegian coast, would cause mass mortalities in salmon net-pens as well as in wild salmon stocks migrating at the time. Over 130 pen-rearing operations in southwestern Norway were moved to safety in the inner parts of the sheltered fjords. The bloom led to the deaths of 500 tons of salmon and rainbow trout, which was a smaller loss level than anticipated.[415] The incident is a serious warning sign with respect to the effects of urban and agricultural pollution in the enclosed Danish seas. It also warns of the possible consequences of contaminating enclosed bays and fjords with food and fecal material from net-pen culture for prolonged periods.

A spiny diatom, *Chaetocerus* spp., has caused problems for Atlantic salmon producers in British Columbia, casting doubts on the suitability of the species in that area.[394] Losses due to algal blooms have also recently been reported from land-based farms.

Oil spills have impacted various regions of the world, with particularly devastating consequence in some instances. Oil spills are of greater concern with respect to their impacts on mammals and birds than fish, as the oil, or at least some portion of it, floats on the sea-surface. Feeding and jumping salmon could be affected and harvesting of salmon would be negated as long as oil was present in the area.

2. Climatic Factors

Overall climate and short-term changes in weather are among the most important external factors to be considered in a net-pen rearing operation. It has previously been pointed out that pens in the British isles, the Faroes, and Iceland have to be much sturdier than those used in Norwegian salmon farming operations. Most Norwegian operations are located inside deep fjords, surrounded by tall mountains, which shelter them from the North Sea storms. Many fjords and lochs in Scotland, Ireland, the Faroes, Iceland, and northern Norway are relatively exposed and storm damage is common, especially in the wintertime. Many sites in Ireland are open to oceanic swells, demanding pens that can withstand a great deal of stress.

Extreme low temperatures that can occur in seawater exposed to harsh winter conditions is uncommon in the middle part of Norway, but is a limiting factor in northern and southern

Norway. Very low temperatures also frequently occur at many Icelandic sites and limit operations to relatively few sites.

Land-based operations, which depend upon electricity for steady pumping, can also be affected by the local climate. Electrical problems due to storms are common in some locations such as Iceland, requiring stable back-up electricity. Large outdoor ponds have a lot of surface cooling in cold and windy weather, lowering the rearing temperature and retarding growth. Seawater pumping is very difficult and costly in exposed coastal areas.

3. Predation

Predation by seals and birds is a serious problem in some net-pen farming areas. Diving birds and seals can make holes in the net and feed on the fish from the outside. Many farms use 10-cm mesh antipredator nets to protect the pens. Seals are one of the biggest threats to salmon farming in Scotland, where grey seals are abundant.[370]

In addition to threats from animals, salmon pens are sometimes exposed to poaching and vandalism, which may require extensive protection measures.

4. Technical Problems

In the same way that net-pen rearing operations are dependent on environmental and climatic conditions, land-based pump-ashore farms are exposed to technical problems. The electric pumps are required to run 24 hours a day and a mechanical breakdown of just one pump can have disastrous consequences. One or more standby pumps is thus a necessity. In addition to mechanical failure, pumps can become clogged with seaweeds and electricity can be unstable. Standby generators usually work properly and can be relied upon to take over in the latter situation, but even then automatic start-up systems can fail with resulting disaster.

Modern facilities are usually equipped with computerized alarm systems which are supposed to react to an emergency in various ways. Experience has shown, however, that blind faith in these types of apparatus is hazardous. The result is that a modern pump-ashore rearing facility has to be much like a modern airplane with respect to having a number of redundant systems, all of which are required on top of the immense construction costs. This shows that pump-ashore farms can only be justified if efficiency and turnover rates are higher and running costs lower than in less expensive net-pen rearing facilities.

C. ENVIRONMENTAL CONSEQUENCES

Like any other industry, aquaculture has its sinister sides. There are enormous waste products, both food remains and fecal material, discharged into the environment from aquaculture production sites. In the case of net-pen culture, much of the material may fall to the bottom of the enclosure or to the sea bottom where it can have negative impacts on future use of the site for rearing. A concern has been expressed in some quarters that the use of antibiotics in the feed used by aquaculturists may impact the natural bacterial flora. There are major concerns regarding the escapees from salmon pens, which may survive in the wild and migrate into salmon streams where they can mate with fish from local populations. Finally, there have been reported incidences of disease and parasite transfer between countries and continents which have had serious environmental consequences.

1. Organic Pollution

Norway is by far the largest producer of farmed Atlantic salmon, producing over 100,000 tons in 1989.[415] This development has raised a lot of questions regarding the impact of salmon farming on the marine environment.

It has been estimated that the salmon farming industry in Norway released over 1,500 tons of phosphorus, 9,000 tons of nitrogen in dissolved form, and 90,000 tons of particulate organic materials into the Norwegian fjords north of Stavanger in 1988.[416] This discharge of organic salts corresponded to 2 million person equivalents. As these discharges are distributed over a large coastal stretch, major eutrophication on a national scale has not been demonstrated but many enclosed local areas are badly polluted.

The organic material frequently collects underneath the salmon pens and decays, at least partly anaerobically, leading to the production of hydrogen sulfide and other poisonous gases, which can cause mortalities in the pens, if they escape into the seawater.[369] Salmon pens are therefore frequently moved from one location to another on a regular basis.

2. Toxins and Insecticides

It has been estimated that the Norwegian salmon culture industry discharged 400 tonnes of impregnating compounds containing copper and tin and 8 tons of insecticides (Neguvon and Nuvan) used against salmon louse in 1986.[416] These chemicals can poison various marine organisms and may accumulate in animals destined for human consumption.

3. Antibiotics

The use of antibiotics in Norwegian salmon culture in 1988 was about 32 tons, corresponding roughly to half a kilogram for each ton of salmon produced in that year. This quantity equals all the antibiotics used for humans and domestic animals in Norway during the same year.[415] This is considered an alarming trend and steps are being taken to reduce the application.

The discharge of antibiotics into the environment can lead to considerable concentrations of the chemicals in various types of fish and shellfish. When exposed fish are eaten by humans, the antibiotics can cause allergies and lead to the development of human pathogens resistant to such chemicals.[416] The potential for development of antibiotic resistant fish pathogens is also of concern.[408]

Antibiotics are also of concern in the bottom mud underneath the net-pens, as the chemicals may affect normal bacterial breakdown and encourage anaerobic processes.

4. Transfer of Diseases and Parasites

It is well known in animal husbandry that the movement of animals or animal products between countries and continents can transfer diseases and other pathogens, which can be of greater concern in the recipient than in the donor country. A classic example is the importation of Scottish sheep into Iceland in the early part of this century. Diseases were introduced which, although relatively harmless to the Scottish sheep, were virulent to local sheep populations.

There are numerous similar examples associated with aquaculture and only a few can be cited here. The best example with respect to pathogenic bacteria is the transfer of the furunculosis organism, *Aeromonas salmonicida,* with live smolts from Scotland to Norway in the mid-1980s. The bacterium had been found earlier in regional rainbow trout culture in Norway. The first major outbreak in net-pens occurred in 1985 on fish farms which had imported smolts from Scotland.[407] Since that time the disease has been of major concern in net-pens in that area.

Viral diseases have also been transferred between continents and recently the disease IHN was reported in southern Europe, presumably imported from the west coast of America with eyed rainbow trout eggs.[410] This virus is common in rainbow trout and Pacific salmon, where it can cause heavy losses.

One alarming example of parasite transfer is the introduction of the fluke, *Gyrodactylus salaris,* into Norway, presumably with smolts from the Baltic region.[411] Experiments have shown that Baltic salmon stocks are resistant to the parasite, whereas Norwegian stocks are not. The parasite, which is found on juveniles in freshwater, is only of minor importance in aquaculture operations, but has spread to numerous salmon rivers and caused a serious decline or elimination of wild stocks. Its presence has been substantiated in 33 out of 220 Norwegian salmon streams. Direct production losses in 1984 were estimated to be equivalent to 250 to 500 tons of adult salmon.[411] Similar differences in susceptibility to the parasite, *Ceratomyxa shasta,* have been reported between stocks of various Pacific salmon species depending on their geographical origin.[417]

One further drastic example of parasite transfer is the introduction of the air-bladder parasite of the eel, *Anguillicola,* from Japan to Europe. *Anguillicola* is common in wild eel populations in Japan, but was nonexistent in the Europen eel. The parasite was first found in eels in Italy in the early 1980s but has since then spread all over Europe.[418] There has been a steady trade of glass eels for culture from Europe to Japan for many years and transfer of live eels back to the European fish markets, which is the most likely path of infection.

5. Genetic Problems

Increased rearing and ranching coincident with overharvest of wild salmon stocks in many areas has aroused concerns regarding genetic integrity of wild populations. On the west coast of North America the focus is primarily on the effects of ranched fish on wild salmon or steelhead populations. Some scientists and managers claim that the present hatchery system, which provides a large share of catchable Pacific salmon, is inefficient and potentially harmful to the remaining wild stocks.[276,419] These issues were also hotly debated in association with the development of private coho salmon ranching in Oregon. An assessment group concluded that the genetic effects of present salmon harvest management were much more profound than any implications of ranching operations.[420]

In Europe, the focus is on the effects of net-pen cultured Atlantic salmon on wild salmon populations. This problem is particularly acute in Norway, where wild salmon stocks have been overharvested for a number of years and the number of escapees from salmon pens has increased drastically. In 1988, it was estimated that the proportion of cultured salmon in the Norwegian salmon fishery was about 20%.[411] Similar problems have been emerging in Scotland, Ireland, and Iceland, where salmon culture is increasing. In the Norwegian situation the cultured salmon have been domesticated and selectively bred, possibly removing some of their adaptive traits for successful survival in the wild. Some geneticists believe that continual flooding of a watershed with fish from exogenous gene pools may seriously retard or even prevent future adaption to the local environment.[421]

There is by no means a concensus on the consequences of genetic mixing of wild and cultured salmon. Some experts claim that the cultured salmon stocks, in spite of selection, have lost no genetic variability and should thus be able to adapt to new environmental situations.[422,423] They also claim that domesticated salmon should be poorly adapted to compete with locally adapted wild salmon. There is also some doubt whether interbreeding of wild and cultured salmon would be very extensive, due to different behavioral patterns in migration, aggression, and spawning.[423,424]

Most scientists agree, however, that many questions regarding this complex problem remain unanswered. Additional research is needed, but in the meantime a conservative approach is advisable as genetic resources, once lost, can not be reclaimed. In response to this problem some countries, such as Norway and Iceland, have resorted to deep freezing genetic material from possibly endangered stocks, and in Iceland a regulatory measure has been established which defines salmon culture policy and sets distances of aquaculture operations from major salmon streams.

VII. PROGRESS IN SALMON FARMING

The previous section describes many of the diseases and problems associated with or facing salmon aquaculture today. Any industry has technical problems, especially if it involves production of living organisms. Considerable research has been ongoing to find solutions to these problems and in many cases satisfactory solutions already exist. There is no doubt, that compared to the pioneering years of salmon culture in the early 1970s, the Atlantic salmon culture industry has overcome many of its most critical problems.

Vaccinations already exist for many diseases and treatments and prophylactic measures have been developed to deal with others. The quality of fish feeds has increased enormously as shown by increased food conversion efficiency and fewer incidences of nutritionally related diseases. Dry feeds have taken over a leading role in the production of foodfish. There has also been a great deal of progress in the design of pens for various environmental conditions.

Selective breeding in Norway is a good example of major progress in a short period of time. Because of selective breeding, maturity of salmon has been delayed and the time for effective growth during the seawater phase increased. This has resulted in the shortening of the time required to produce a 4 to 5-kilogram average size fish by as much as 6 months in only 8 years.[425] Similar progress has taken place in the rearing of smolts, resulting in increased smolt quality. Continued progress will certainly increase the efficiency of the Norwegian salmon rearing industry and make it more competitive even during difficult periods.

We will undoubtedly see continued progress in the development of salmon farming and satisfactory solutions to most of the technical and biotechnical problems. Due to the exposed nature of net-pen farming in particular, the environmental issues may be more difficult to deal with, and economic as well as socioeconomic issues will continue to be of great concern.

VIII. DEVELOPMENT IN ATLANTIC SALMON PRODUCTION

This section reviews the status of Atlantic salmon culture in various countries and includes some production statistics. The countries are discussed in order of decreasing importance with respect to salmon culture. A brief section describes the major rearing methods used in each country, how they relate to local environmental conditions, and the main factors restricting production.

Figure 31 shows the production of cultured salmon in various countries over a period of 10 years. It is clear that Norway, with a 1989 production of 115,000 tons, is producing four times as much as its closest competitor, Scotland, with a production of 28,000 tons. The Faroe Islands and Ireland are approaching 8,000 tons, followed by Eastern Canada and Iceland, which are well below 4,000 tons. Production of Atlantic salmon in Chile reached 1,000 tons in 1989, but in the United States and western Canada the production of Atlantics has, in many instances, been experimental, and combined production has been well below 1,000 tons.

The total world production of farmed Atlantic salmon in 1989 was thus close to 164,000 tons, which is almost 20 times the average annual wild harvest of the species. It is interesting to note that this quantity of farmed Atlantic salmon probably exceeds the world harvest of chinook and cohos, its prime competitors, by a wide margin, demonstrating the economic impact of the industry in various parts of the world.

Although not covered in this chapter, it should be noted that there is a large production of rainbow trout in the European countries, some of which is directly competing with Atlantic salmon in the marketplace. Total trout production in 1986 was in excess of 90,000 tons, of

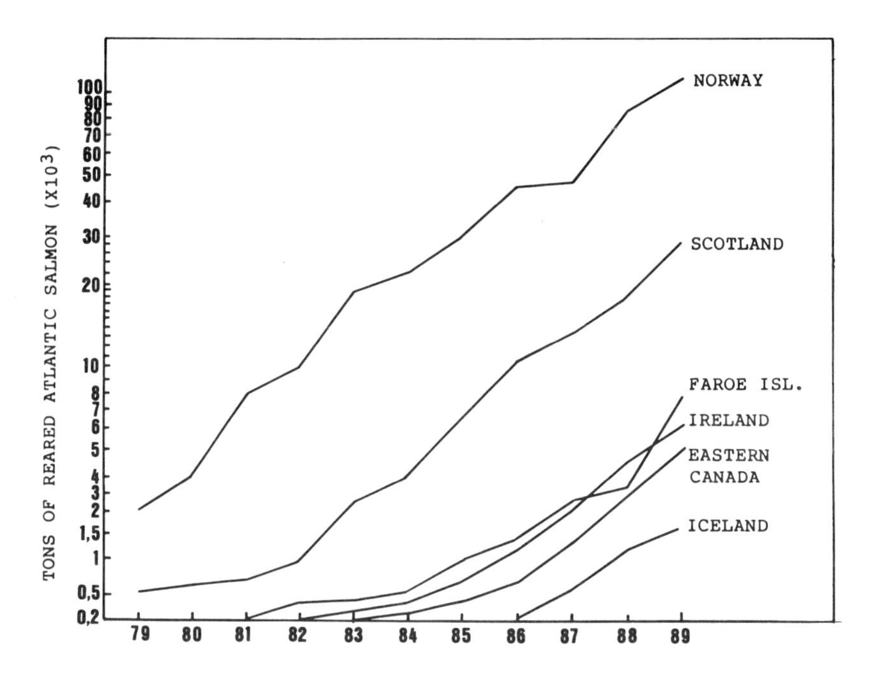

FIGURE 31. Production of farmed Atlantic salmon in various countries on both sides of the Atlantic Ocean. Production in the countries listed is already 20 times the combined production of wild and ranched Atlantic salmon.

which 80% was portion-size trout from Denmark, France, Germany, and Spain.[426] The remainder was 2- to 4-kilogram trout from Norway, Sweden, and Finland.

A. NORWAY

Norway was the first country to embark on large-scale net-pen culture of Atlantic salmon in the early 1970s. Initial industry growth was slow and had, by 1980, only reached 4,000 tons. After that the expansion was dramatic, reaching 30,000 tons in 1985 and exceeding 100,000 tons in 1989 (Figure 31). In 1989, there were over 700 farms in Norway producing about 115,000 tons of salmon. Production in 1990 was estimated at 160,000 tons.[427]

Norway is especially well suited for net-pen culture of salmon with its sheltered, deep fjords and tempered seawater. Over 90% of the production comes from the areas north of Stavanger and south of Bod. There are ample sites for increased net-pen production in Norway, and production has in the past primarily been limited by smolt supply and size limitations imposed by law.

In the 1970s there was a chronic shortage of smolts in Norway. Smolts were imported from Sweden, Finland, the United Kingdom, and Iceland to try to satisfy the demands of the rapidly growing industry. As previously pointed out, those imports have in some instances had serious ecological consequences. In the 1980s, new smolt production facilities were developed and by 1987 Norwegian production had reached 45 to 50 million smolts, an overproduction of about 13 million.[428] This develoment has already increased demand for larger smolts at similar or lower prices,[429] and there has been considerable speculation regarding ocean ranching of excess smolts. These ideas have been of considerable ecological concern as many Norwegian rivers already have large numbers of salmon strays from pens, which in some cases outnumber wild salmon returning to the rivers. As a result, the Ministry of the Environment has decided not to allow ranching of excess smolts from the Norwegian smolt farming industry.

The Norwegian salmon farming industry has been regulated by the central government since the enactment of fish farming legislation in 1973. The primary goal was to manage the growth of the industry while favoring its development in rural areas. It was considered undesirable to have a large share of the production in the hands of large national or multinational corporations. To achieve these ends, all new farms had to acquire permits and were limited to 3,000 cubic meters of rearing volume until 1981, when the limit was raised to 8,000 cubic meters, corresponding to roughly 150 to 200 tons of salmon production.[426] From 1977 through 1981, no new farming licenses were issued, primarily due to the fact that smolt production was not keeping pace with the rapidly expanding net-pen culture industry.[430]

The existing regulations have usefully served the purpose of creating a rural industry, but the side effects have included a growing export of Norwegian capital and expertise to other countries. Increased experience and progress have demonstrated that the government restrictions were also preventing the economies of scale in the size of individual farms in Norway.[362,431] This also prevented the operation of multiple sites and the vertical integration of smolt production. As several studies have shown, considerably larger fish farms would be more economical to run, so restrictions might place the Norwegian salmon farming industry at a severe competitive disadvantage when prices are down.[362]

In addition to economic drawbacks, the nature of the restrictions may have forced fish farmers into questionable rearing practices. High salmon densities have undoubtedly created a precarious and stressful situation for the fish when farmers have attempted to maximize production without increased net-pen volume. Many fish farmers have been skeptical of the effectiveness and fairness of the volume regulations, considering that the applicable fish densities in net-pens vary between localities and the danger of disease outbreaks increases with higher densities.[432] The 1973 legislation was revised in 1981 and again in 1985.

There are no direct subsidies given to the aquaculture sector in Norway, but many of the fish farms in remote districts receive a 10 to 35% investment grant for capital investment and a state guarantee for working capital requirements.[426,430] The Norwegian aquaculture industry operates under the auspices of the Ministry of Fisheries.

In 1970 the salmon farmers founded the Norwegian Fish Farmers Association, which functions as a spokesman for the industry in negotiating with the government in shaping the framework of the industry.[361] The Norwegian Fish Farmers Sales Organization is also owned by the fish farmers and controls the sales and quality of all farmed fish in Norway, a right protected by law. Fish farmers can only sell their product through buyers recognized by the sales organization. Individual farmers, on the other hand, have the benefits of being able to sell all their annual production at a price which does not fall below a certain level. This has led to overproduction of salmon in recent years.

The export of farmed Norwegian salmon started in the early 1980s and was mostly focusing on England and continental Europe.[361] Exports to the United States began in 1981 and continued to rise through 1986, when the United States was importing almost 25% of the total Norwegian production, matching imports to France. Those two countries continue to be the largest customers for Norwegian cultured salmon.[415] Exports to Japan have also been increasing.

There are over 40 institutions involved in aquaculture research in Norway, but there are four primary ones which conduct research in all aspects of the industry.[415] The Marine Institute in Bergen with its field stations at Matre and Austevoll has been instrumental in the technological development of the industry and disease research. Many other Bergen institutes are involved as well as the University of Bergen.

The Veterinary Institute in Oslo does research in fish diseases and has disease inspection and management responsibilities. Akvaforsk at the Agricultural University in Aas conducts aquaculture research and selective breeding at its field stations at Sunndalsra and Averoy. In addition, there is a significant amount of aquaculture research ongoing at the University of Tromso, the Norwegian Institute of Nature Research in Trondheim and various other institutions in the Trondheim area.

B. SCOTLAND

Salmon farming in Scotland started only a few years after that in Norway, but initial growth was comparatively slow.[364] Unlike Norway, where the industry was envisaged as providing added income for rural economy, the industry in Scotland has been largely developed by large national or multinational companies. Nevertheless, rural economies have benefited from the employment; e.g., in Shetland, farms are owned locally as in Norway. Marine Harvest, a subsidiary of Unilever, built the first salmon farm in Scotland in 1966 and is by far that country's largest producer.

Although growth was slow in the beginning, the Scottish industry has, since 1982, been growing at a similar rate to the one in Norway. Scotland is the second largest producer of cultured salmon in the world, producing over 28,000 tons in 1989 (Figure 31). Over half of the production is in the hands of three large companies.[433] In 1987 there were 126 companies producing salmon at 196 net-pen sites and 11 land-based pump-ashore sites.[434]

Most of the salmon pens are located in sheltered sea lochs along the west coast of mainland Scotland, the Hebrides, Orkney, and Shetland. The conditions are less suitable than in Norway and net-pens are, on the average, smaller and stronger. Mainland Scotland now suffers from a shortage of suitable sheltered sites and new farms will have to use more exposed sites at greater cost.[364] Some experts believe that Shetland may eventually produce more salmon than mainland Scotland, as it has more sites, some of which are very well sheltered.[435] Most pump-ashore facilities in Scotland have had financial difficulties in recent years and their future sems somewhat uncertain.[436]

As in Norway, the Scottish salmon farming industry originally had to endure a smolt shortage. Ironically, in the early 1980s some of the smolts produced in Scotland were exported to Norway in spite of local shortages, probably due to heavy Norwegian interests in many Scottish enterprises. In 1989, the total production of smolts in Scotland had reached 26 million, with a considerable surplus over local need.[436] Unlike Norway, Scotland has had a conservative policy regarding smolt imports, which have been entirely forbidden.

As in many other salmon farming countries, there are some concerns regarding the intrusion of escaped salmon from net-pens into Scottish rivers. Unlike Norway, however, most of the major salmon-producing river systems are on Scotland's east coast, whereas most of the salmon farms are on the more sheltered west coast.

The Scottish industry is somewhat less effective than the Norwegian one, due to inferior sites, smaller and more expensive net-pens, as well as more storm damage. These differences are demonstrated by the fact that the Scottish industry uses 650 smolts to produce a ton of salmon, vs. 350 in Norway.[433] This is, however, partly offset by lower smolt production costs in Scotland, where great numbers are produced in freshwater cages.[436] Additional problems are caused by early sexual maturation.

In mainland Scotland and the Hebrides almost 20% of the standing stock is harvested as grilse in early summer, at an average weight of 2.1 kg.[436] In Orkney and Shetland the maturation pattern is similar to that in Norway and most of the salmon are harvested in early winter when they are over 3 kg in average weight.[364]

The Scottish aquaculture industry has had very little government control, but considerable public support has been channeled to the industry from various sources, e.g., through the Highland & Islands Development Board. Between 1965 and 1983 the Board allocated 11.7 million British pounds to fish farming, of which 3.5 million pounds was for research and development.[437] Banks have, on the other hand, been less willing to lend money to salmon farmers than has been the case in Norway, which partly accounts for slower expansion and healthier finance structure in Scotland.[433] Administrative responsibilities rest with the various departments of the Ministry of Agriculture and Fisheries for Scotland and with local government agencies.

Due perhaps to the structure of the industry, with a few companies dominating the scene, Scottish fish farmers do not have a central sales organization. They must fend for themselves in the marketplace and be on the alert for overproduction. To their benefit the fish farmers have a large local market within the United Kingdom and great quantities of farmed salmon are sold directly to smokers, various fish markets such as Billingsgate in London, or directly to retail or catering outlets, where the product must compete directly with wild Atlantic salmon, frozen Pacific salmon, and fresh imported farmed fish.[364] Exports have risen sharply since 1984, mostly to other EEC countries.[426] The Scottish Salmon Growers Association lays down specifications for harvesting procedures and product quality for all Scottish farmed salmon through the Scottish Salmon Board, a daughter organization.

Several research institutions and universities in Scotland carry out research in aquaculture and fish diseases. Included are the Marine Institute in Aberdeen, the Freshwater Laboratory in Pitlochry, and the University of Sterling.

C. FAROE ISLANDS

The Faroe Islands rank third as a major producer of cultured Atlantic salmon, with an annual production of 8,000 tons in 1989 (Figure 31). The islands, which are located in the Atlantic Ocean west of Bergen, Norway, have ideal temperatures for salmon rearing, ranging from 5°C in winter to 10°C in summer.[438] Sheltered sites, however, are limited, which has led to crowding of net-pens in a limited area with corresponding disease problems.

Salmon farming was started in the Faroes on an experimental basis in 1973 after the establishment of the research facility at Air, but was only of small scale until 1984, when it started to increase dramatically (Figure 31). Shortage of smolts has slowed the growth of the industry, and for many years the only supplier of smolts was the research station at Air. At the present time there are 17 smolt stations, with a production capacity of seven million smolts, and over 60 net-pen rearing sites in operation in the Faroes.[439]

Due to the shortage of sheltered sites in the Faroes the fish farmers have had to crowd their pens into a limited area with some environmental and disease consequences. It seems likely that further expansion will depend on the use of strong, high seas net-pens. Several such pens of the Bridgestone type are already in use.

Since 1978 the Faroese salmon farmers have exclusively been using locally cultured Norwegian salmon stocks and Norwegian technology can in most cases be directly tranferred. There are only minor introduced wild salmon runs on the islands, which obviate ecological concerns regarding imports of foreign stocks. Concern does exist, however, regarding past introductions of IPN viral disease with salmon eggs from Norway.[440]

The location of the Faroe Islands close to the major foraging areas of wild Atlantic salmon has created interest in salmon ranching. Experiments with local and Norwegian stocks have given encouraging results.[438]

Salmon culture in the Faroes has evolved without government support. Fish farmers have, however, been able to obtain loans for capital investment and running costs from industrial and developmental funds as well as from the Faroese fisheries sales organization.[441]

Fish farming in the Faroes is not designated to any one ministry, but is usually dealt with by the ministries of fisheries, industry, or agriculture. Administrative and research responsibilities rest with the Fisheries Laboratory in Torshavn, using the research facilities of p.f. Fiskaaling, a government enterprise which operates several fish farming sites in the Faroes.

Although salmon farming is getting to be of great economic importance in the Faroes, profitability in the industry has been marginal.[442]

D. IRELAND

Ireland has a long history of salmon enhancement, primarily with eggs and fry.[443] Large-scale salmon culture was first undertaken by the Electricity Supply Board with the aim of restocking rivers affected by hydroelectric plants. By 1970, smolt production methods were well developed through the pioneering work of the Salmon Research Trust experimental facility at Furnace in Newport.[356] In recent years public releases have been in the order of half a million smolts annually.[444]

As many other European countries, Ireland started experimenting with net-pen rearing of salmon in the early 1970s. Initial trials in Killarny harbor on Ireland's west coast showed that pens of classical Norwegian design were not suitable for the exposed Irish coast.[367] Some experiments in more sheltered areas were initially promising, but mass mortalities due to excessive summer heat in 1983 and 1984 demonstrated that Ireland was at the southern border for successful inshore farming and should go for offshore technology.

In early 1985 several Irish companies decided to buy strong net-pens for exposed areas and by 1989 there were over 300 units of the Irish designed "Wavemasters" used in various areas.[445] Another breakthrough was the introduction of the Bridgestone net-pens into Ireland in 1983. These pens, basically giant rubber hexagons with over 100-ton capacity, have opened up a vast range of exposed sites for Irish fish farmers.[367] There are about 30 such units in use in Ireland today.

In 1989 there were some 34 marine farming sites in operation in Ireland, producing in excess of 5,000 tons.[446] Production has increased to that amount from a mere 700 tons in only 4 years (Figure 31). The three largest fish farming companies, which have multiple sites, account for over 50% of the production. Most of the production in Ireland is exported to nearby France, which seems to have a preference for the relatively small Irish salmon (1 to 3 kg). Several Irish companies are also looking into the possibilities of selling processed salmon to other European countries.[447] Being a member of EEC, Ireland has the advantage of being able to sell its products without protective taxation.

In the 1980s there has been a great increase in the commercial production of smolts for the aquaculture industry. The 1985 output was almost 600,000 smolts.[426] In the early 1980s there was a serious smolt shortage and smolts were imported from Norway and Iceland under strict quarantine conditions. In 1989 there were 25 smolt production units in operation, producing over seven million smolts. There was a slight overproduction and smolts were exported for the first time.[447] Early maturation has been a serious problem in Irish salmon culture, but has decreased in some farms after importation of selectively bred Norwegian stocks.

In 1988 the Irish Salmon Growers Association was registered as a company. It is funded by a membership fee and a levy on smolt and salmon sales. The organization represents Irish salmon farmers and keeps track of quality, markets, and research.[445] A new BP fish food factory, which opened in 1989, released the Irish market from total dependence on United Kingdom feed plants.[447]

The Department of the Marine is responsible for the formulation and implementation of policies related to aquaculture. It provides services from its aquaculture and engineering sections as well as biological and disease research and advice from the Fisheries Research Center in Dublin.[445] The Central Fisheries Board, various local boards, as well as the universities in Galway and Cork, provide advice to fish farmers.[445]

Financial assistance is available to Irish Fish Farmers from BIM (Board Iascaigh Mhara) and UDNG (Udaras na Gaeltachta) in combination with EEC grants.[445] Direct subsidies may be awarded to pilot projects at levels up to 50% of fixed assets, and commercial projects may receive up to 10% of capital costs.[426]

E. EASTERN CANADA

The salmon farming industry in eastern Canada is concentrated in the Bay of Fundy in New Brunswick and to a lesser extent in Nova Scotia. Initial experiments to grow salmon in those areas in the early 1970s were unsuccessful due to extremely low seawater temperatures in the winter.[448] Salmon were first successfully grown in 1979 at Deer Island in the outer Bay of Fundy and production has since then grown from 11 tons in 1980 to 4,000 tons in 1989 (Figure 31).

In 1988 there were 34 farms operating in the Bay of Fundy, producing over 3,000 tons, while 13 sites in Nova Scotia were estimated to produce over 500 tons. Salmon farming in Quebec is limited to land-based facilities due to severe winter conditions and salmon culture in Newfoundland must be considered experimental.[449]

Land-based seawater aquaculture is actually not a recent development in Canada, as such a facility "Sea Pool Ltd." was operating in Nova Scotia in the early 1970s. It is noteworthy that there has been some interest in constructing land-based facilities in areas where net-pen culture is possible, such as in British Columbia and Prince Edward Island.

The major limiting factor facing the salmon farming industry in eastern Canada is limited availability of suitable sites. Most of the coastal Atlantic regions have proved too cold and some net-pen sites may be suitable for rearing in most years, but have occasional lethal winter temperatures. It has been estimated that approximately 75 sites may be suitable for salmon farming, of which more than 35 are already being utilized.[449]

Smolt supply has restricted the growth of the salmon farming industry in New Brunswick and Newfoundland and will probably continue to do so in the near future. In 1987 about 1 million smolts were supplied to the aquaculture industry in New Brunswick from one federal smolt station and 11 private stations.[448]

As in many other countries, there are substantial environmental concerns regarding straying of salmon from net-pens into rivers and with respect to pollution from the pens. These concerns exist primarily in the Bay of Fundy, where most of the industry is concentrated. Environmental regulations are, however, not expected to limit the growth of the industry.[449]

Canadian federal and provincial governments share jurisdiction over the aquaculture industry. Both levels of government have agreed that the lead development role for aquaculture rests with the private sector. The governments provide the appropriate investment climate and regulatory framework. The Canadian federal government grants aquaculturists assistance through general programs available to a wide range of businesses. Provincial support occurs via programs for industrial development, intraprovincial marketing, and training.[426] There has been concern regarding lack of available financing for salmon farms in the start-up years.[449] The government of New Brunswick has promoted medium-sized farms and local ownership, although there are a few large integrated operations in existence.[448]

There are number of research facilities and institutes which serve the aquaculture industry in Canada. An aquaculture demonstration and development farm under the supervision of the Department of Fisheries and Oceans was opened at Lime Kiln Bay in 1986. The purpose of that facility is to facilitate the transfer of net-pen culture technology to the developing aquaculture industry. The St. Andrews Biological station, the Atlantic Salmon Federation research facility, and the Huntsman Marine laboratory have all been instrumental in the development of the aquaculture industry.

F. ICELAND

Iceland has a long heritage in salmon enhancement. Although the first salmon hatchery was built in Iceland in the 1880s, 50 years were to pass before an efficient hatchery for incubation of salmon ova was established. Only unfed fry were released into rivers until 1950, but at that time a few hatcheries started rearing fry for a few months before releases.[450] In the 1960s the state-owned Kollafjördur Experimental Fish Farm started rearing salmon to the smolt stage for releases into rivers and for ocean ranching.[386]

In the last 20 years many aquaculture enterprises have started producing food fish, with most of them coming into production in the 1980s. Many of the newest salmon farms have considerable foreign investment, primarily from Norway and Sweden. The only restrictions applied to salmon farms in Iceland are related to environmental concerns. Foreign investment has been encouraged, but foreign investors can not own more than a 49% share in any business. The trend in Iceland has been towards large land-based salmon farms and salmon ranching ventures. Most of these concerns have chosen to build their own smolt rearing facilities to ensure supply.

In 1983, there were 30 rearing stations, chiefly working with Atlantic salmon, but to a limited extent with rainbow trout, Arctic char, or brown trout. They produced less than 50 tons of food fish and 500,000 salmon smolts. By 1987 there were over 100 fish farms producing over 500 tons and in 1989 production was in excess of 1, 500 tons (Figure 31).[451] The primary increase in production has thus taken place in the last 3 years. Smolt production has similarly increased from 200 thousand in 1980 to 10 million in 1990, a fifty-fold increase. This production, being considerably in excess of the present requirement of smolts for rearing and ranching, is of great concern to the smolt producers.

Conditions in Iceland are not well suited for conventional net-pen rearing of salmon. Sheltered areas are limited, winter storms frequent, and undercooling of seawater common in some areas. One of the largest net-pen rearing operations is on a rather unique brackish lagoon on the north coast, but otherwise most of the net-pen rearing is concentrated in the southwest, close to the capital city of Reykjavik, where sturdy oceanic net-pens are dominating the scene, such as Bridgestone, Farmocean, and Wavemaster. The success of the large oceanic pens is still largely uncertain. More net-pens are being situated on Iceland's east coast, where average seawater temperatures are lower, but temperature fluctuations in the winter are less pronounced.

Many land-based farms have been built in Iceland, primarily in the southwest, where many stations harness thermal energy to enhance growth. These farms account for over half of the present production.[451] Due to the large investment costs and lower salmon prices, most of the land-based operations encountered financial difficulties during 1989 and many went bankrupt in 1990, although they remain in operation.

Iceland has relatively favorable conditions for smolt production, having ample ground water resources. The well water, being only 4°C, must be heated to favorable rearing temperatures using geothermal resources. Given the fact that the oceanic fishery for salmon is forbidden within the 200-mile economic zone, many fish culturists believe that ocean ranching may be an economical way of producing salmon which will not compete directly with cultured Atlantic salmon. The production of ocean ranched salmon already exceeds that of the local wild salmon populations. Due to unfavorable oceanic conditions from 1988 to 1990, return rates have been low and many of the larger ocean ranching facilities have encountered serious financial difficulties.

The recent expansion in salmon aquaculture has aroused considerable concern regarding the genetic integrity of wild salmon. Some of the net-pen rearing operations are located in the vicinity of good salmon rivers and considerable straying of escapees into the rivers has been observed. In 1988 a regulatory measure was set to minimize these interactions.[452] Some of the major provisions included the setting of minimum distances from river estuaries to net-pens and ranching stations. It was also forbidden to rear foreign stocks in net-pens. Norwegian salmon eggs had previously been imported to Iceland under close supervision of Icelandic veterinarian authorities. When the eggs had been successfully quarantined and were ready for further distribution, it was considered unwise to face the risk of getting those stocks mixed with local wild populations and distribution was limited to land-based facilities.

In 1986 the Icelandic government, possibly in response to the interest of local and foreign investors, decided to place major emphasis on the development of salmon aquaculture in Iceland. Research funds were increased considerably and significant numbers of investment loans were granted to Icelandic investors for building of rearing stations in various areas of the country, mostly from the Development Fund of Iceland. Other financial institutions were, however, reluctant to lend money to aquaculture enterprises and most stations had major difficulties in financing their operating costs. Finally, a partial government guarantee was enacted to support loans taken to cover initial operating costs. Nevertheless, as a result of reduced salmon prices and a major setback in the economics of salmon rearing in 1989, many Icelandic salmon rearing stations have experienced great financial difficulty.

Farming of salmonids is under the auspices of the Ministry of Agriculture, assisted by the Institute of Freshwater Fisheries, whereas rearing of oceanic fish is under the Ministry of Fisheries. Several research institutions are involved in different aspects of aquaculture research, including the Institute of Freshwater Fisheries, various departments of the University of Iceland, the Agricultural Research Institute, the Institute of Marine Research, and various private companies.

G. PRODUCTION IN THE PACIFIC

Salmon farming has been growing rapidly in many regions bordering the Pacific, such as Japan; British Columbia, Canada; Washington, U.S.; Chile; New Zealand; and Tasmania. The primary species produced in these areas are Pacific salmon, primarily chinook (*Oncorhynchus tshawytscha*) and coho (*Oncorhynchus kisutch*). In 1988 the total production of these species was about 30,000 metric tons with Japan and British Columbia leading with 15,000 and 8,000 metric tons, respectively.[431]

It is generally accepted that the Pacific salmon species are more difficult to grow to market size than rainbow trout and Atlantic salmon. Pacific salmon are more prone to disease and experience in British Columbia has demonstrated that only 5 to 8 kg of Pacific salmon can be raised in a cubic meter of net-pen, as compared with 15 to 20 kg/m^3 for Atlantics.[449] A great deal of interest has been generated in the farming of Atlantic salmon in Washington and British Columbia. The largest coho salmon farm in Washington was recently sold and is shifting its emphasis to Atlantic salmon.[431] Most new permit applications in the State of Washington are also for facilities that plan to rear Atlantic salmon.

Although conditions in British Columbia and Washington seem to match those of Norway with respect to Atlantic salmon culture, success should not be taken for granted. Atlantic salmon reared in British Columbia have been extremely susceptible to blooms of *Chaetocerus* spp., a spiny diatom prevalent in the area.[394] The Atlantic salmon produced in British Columbia have also been hard hit by the myxosporidian *Henneguya* spp., which is not found in areas from which the Atlantic salmon were introduced. Similar suceptibilities have been reported from Atlantic salmon stocks introduced to Tasmania and Chile.[394] The use of local stocks, if at all possible, can thus not be overemphasized. Such stocks of Maine origin have been experimentally cultured at the National Marine Fisheries Service laboratory at Manchester, Washington since the late 1970s, making semi-local stocks available to the industry.

IX. ECONOMICS OF SALMON CULTURE

With the rapid increase of salmon culture in many countries in the late 1980s it was expected that prices would fall drastically at some point. When Norwegian production doubled from 40,000 to 80,000 tons between 1987 and 1988, markets were ready for a price reduction during 1988 which did not materialize due to poor catches of chinook and coho in the Pacific.[453] Considering that there was an increase of over 50,000 tons between 1988 and 1989, the present crisis in the industry should not surprise anyone.

It is highly likely that the industry will be forced to accept lower prices in the future, and the image of salmon must be changed from that of a luxury item to that of a dietary staple. The success of the salmon culture industry in each country will probably depend on the ability of those nations to lower production costs and adjust to the new goal.

Due to the seasonal nature of salmon fisheries, there is no doubt that cultured salmon will continue to replace wild fish in world markets. Cultured salmon are available year round and a steady supply can be guaranteed. Wild salmon are only available during the summer months and, for the most part, they are not bled and processed according to the specifications of exacting customers.

Ranched Atlantic salmon, which are only available at this time in Iceland, have many things in common with the wild populations, but are in most cases processed according to strict specifications. Many salmon ranchers believe that ranched fish will fall in a category of their own and will cater to a specific market seeking a different meat texture, absence of antibiotics, and different size categories. Since the quantities of ranched salmon may never be very large, the producers also speculate that the fish will bring premium prices.

For the time being it seems likely that both the salmon farming and ranching industries will face some very difficult times in the next few years. For several reasons the fast expanding industry in Norway may be hardest hit. Among those reasons are

1. Norway has a very small local market and must rely almost entirely on exports to other European countries, the United States, and Japan.
2. There are more than 70 companies with export licenses, which compete internally and are already having difficulties with profits.[362]
3. With growing aquaculture in the United States and Canada, it will become increasingly difficult for Norway to compete in the North American market.[431]
4. Continued size limitations of the industry will reduce cost effectiveness and prevent competitive pricing.[362]
5. Continued expansion may require limitation in total production.

It should, however, be pointed out that Norway has several factors in its favor, such as some of the best natural conditions for relatively inexpensive net-pen culture, a well established feed industry, and successful breeding programs. These characteristics have increased the profit margin of salmon farmers. There has, furthermore, been a lack of serious conflict over resource development and allocation, which is fairly prominent in North American aquaculture. These factors may, in the long run, offset many of the above-stated disadvantages.

The Norwegian Fish Farming industry is already responding to these problems. In 1989 over 7,000 tons of salmon were frozen and stored instead of being sent to the market.[454] The record 1989 production was further reduced by keeping 20,000 tons of salmon over the winter and into the spring of 1990.[455] The minimum price paid to fish farmers by the Norwegian Sales Organization was also lowered by 15%.[456]

In order to meet falling prices on the world market, all fish farmers must reduce their costs. There are three primary factors that contribute to high production costs: feed, smolt prices, and finance charges.[362] Of these, the most likely candidate for reduction is the cost of smolts. Overproduction of smolts in some countries like Norway and Iceland is a step in that direction.

It is seems likely that land-based farms with high finance charges will be having a hard time, given current economic conditions, unless they find a way to increase densities or use their effluent for some kind of polyculture of shellfish or seaweed.

With the exception of Norway there is a clear trend in most countries towards fewer and larger production units. These corporations should be able to reduce risks through multiple-site operations and to control input requirements in the form of eggs, smolts, and feed through vertical integration.[431] This trend is quite clear in Scotland, Ireland, Canada, and Iceland, where a large share of the production is in the hands of a few large companies.

Chapter 6

COHO SALMON FARMING IN JAPAN

Conrad V.W. Mahnken

TABLE OF CONTENTS

I. INTRODUCTION

Japan is the world's largest consumer of salmon. In 1985, for example, Japan's consumption of 349,000 metric tons (catch plus imports) was 38% of world harvest.[457] Predictable supplies of this important source of quality marine food are considered vital to meeting the protein needs of the island nation. Two factors have provided impetus for the development of salmon ranching and salmon farming in Japan: growing national demand and the establishment of extended jurisdiction to 200 miles by most countries in the 1970s. Enactment of 200-mile exclusive economic zones by most nations in 1973 displaced the traditional Japanese high seas fleets; consequently, the high seas catch, which was the principal source of salmon, dropped from 120 thousand tons in 1965 to less than 25 thousand tons in 1985. Therefore, the Japanese have turned to aquaculture to offset the deficit in balance of trade created by rising imports. In the 1970s federal and prefectural governments, fishing companies, and fishermen cooperatives embarked on what has become the world's most successful sea ranching program for chum salmon (*Oncorhynchus keta*) and more recently have turned to net-pen farming of coho (*O. kisutch*) to further diversify the supply of salmon products.

Japan has a long history of salmon culture. The first propagation experiments were carried out in 1876 and the first national chum salmon hatchery was established at Chitose on Hokkaido 1 year later. The Japanese salmon aquaculture program presently consists of a combined salmon ranching and farming effort. The salmon ranching program in Japan involves hatcheries that release more than two billion chum salmon fry and lesser numbers of pink (*O. gorbuscha*), masu (*O. masou*), and sockeye (*O. nerka*) salmon juveniles which return as adults for harvest in the high seas and coastal fisheries. The intensive marine farming of coho salmon is carried out in net cages.

II. HISTORY OF COHO FARMING

Although the salmon ranching program in Japan has been well established, salmon farming is a relatively new phenomenon. The rise of coho salmon farming in Japan had its origins in the 1960s when Dr. Akimitsu Koganezawa and his associates began experimenting with saltwater culture of rainbow trout (*Oncorhynchus mykiss*), and a small commercial industry developed in 1967. By 1975, large rainbow trout were being produced at farms in Miyagi Prefecture on Ogatsu Bay, with an annual production of 300 tons, but because the local market was unfamiliar with large-size rainbow trout, the farmers could not command a reasonable price. Coho salmon farming technology evolved as a natural outgrowth of marine trout farming. Pilot-scale experimentation with other salmonid species continued in Miyagi and Iwate Prefectures and to the north on Hokkaido. Chum, pink, chinook, sockeye, and coho salmon were all eventually cultured experimentally in net-pens, tanks, or ponds.

The rising consumer demand for salmon and growing national expertise in salmonid culture accelerated interest in domestic salmon farming. The Nichiro Company established a salmon culture division in 1970 and began a series of experiments with native and exotic salmonids.[457] Freshwater culture of nonnative coho, chinook, and sockeye salmon as well as native chum and pink salmon began in 1971 after high seas quotas were imposed and salmon factory ships in Japan became subject to a vessel reduction program. In 1973, eyed coho salmon eggs were imported from public hatcheries in the states of Washington and Oregon and raised to smolt size in a freshwater lake in Shizuoka Prefecture south of Tokyo.

Commercial production began in November 1974 when Nichiro transferred smolts to net-pens near Kurihama (Tokyo Bay) Kanagawa Prefecture in 1974.[458] Encouraged by their success, they began to seek out additional farming environments in the more optimal coastal bays of the Sanriki District, to the north of Tokyo. A farming system began to take shape in 1979 when, after repeated saltwater rearing trials, the company was able to produce 350 tons of coho salmon with 80% survival from smolt to adult.[459] After 4 years of study, the company chose coho salmon as its primary cultivar because of disease resistance, good survival and growth in sea cages, and marketability. The Nichiro Company joined forces with Fisheries Cooperative Associations of Miyagi and Iwate Prefectures, and by 1984, Japanese production increased to 4,400 tons. Nichiro's success did not go unnoticed. In 1979, the industry was entered by another high seas fishing corporation, the Taiyo Fishery Company, and by a manufacturer of fishing gear, the Nichimo Company.

Farming of chinook salmon was also attempted by the Nichiro Company.[457] Eggs were imported from Washington state and juveniles reared for 2 years in freshwater where they attained 350 g average size. Juveniles were shipped to two saltwater farms, one in Miyagi Prefecture and another in Niigata Prefecture. Eighty percent of the fish survived,[460] but despite 1 additional year of freshwater rearing, growth was poor and the chinook only attained the size of coho salmon (2.5 kg).

The Japanese production of pen-farmed coho salmon has increased from 72 tons in 1978 to 16,000 tons in 1988, a more than 200-fold increase in a decade (Table 12). Production failed to increase annually only once, in 1981, as a result of fish losses to a typhoon in December 1980. The entire production currently is consumed domestically; future increases in production are not expected to be exported. Thus, salmon farming may eventually reduce Japan's reliance on capture fishery imports, the majority of which come from the United States. Japan maintained its position as world leader in the production of farmed coho salmon through 1988, but is now being challenged by Chile as its share has dropped to roughly 50% (Table 12).

It is estimated that farmers are now producing at near the maximum level that the domestic markets can bear, and a steady decline in farm revenue from about 1,200 to 900 yen per kilo since 1982 may indicate overproduction.

III. PUBLIC POLICY AND REGULATION OF SALMON AQUACULTURE

A set of regulations called the Fishery Rights System[461] provides the legal framework for salmon aquaculture in Japan. Fishing rights are authorized by the federal government and granted by prefectural governors.[462] The laws consolidate power in the hands of village fishermen and give local autonomy to the fishing cooperatives, the parent organizations to aquaculture cooperatives. The result is an industry dominated by small family businesses. A major factor contributing to the success of aquacultural projects is the ease with which public waters can be obtained for this use.

The three categories of rights are

1. *Common Fishing Rights.* These are Fishing Rights granted to fishermen cooperatives and are renewable every 10 years. They are monopolistic and exclusive rights to engage in the harvest of shellfish, other sedentary animals, and seaweeds in shallow public waters. In theory, the rights must be periodically renewed and can be denied and passed to others at the time of renewal. In practice, rights seldom pass to new ownership unless the applicant no longer desires them, or there is adequate room for expansion of the fishery.
2. *Setnet Fishing Rights.* These apply to net fisheries in waters deeper than 27 m and are operated in a manner similar to Common Fishing Rights. Setnet Fishing Rights are renewable every 5 years.

TABLE 12
Japanese Production of Coho Salmon 1978—1988 (metric tons)

Prefecture	1978	1979	1980	1981	1982	1983	1984	1985	1986	1987	1988	1989 (est.)
Miyagi	72	350	1,855	1,150	1,900	2,600	4,000	6,400	6,800	10,600	14,000	16,500
Iwate	0	0	0	0	200	300	400	500	600	900	1,000	1,200
Niigata	0	0	0	0	0	0	0	100	200	500	700	900
Others	0	0	0	0	0	0	0	0	100	200	300	400
Total	72	350	1,850	1,150	2,100	2,900	4,400	7,000	7,700	12,200	16,000	19,000
Number of farmers	7	39	99	105	112	121	157	167	197	266	n.d.	n.d.
Revenue (yen/kg)	n.d.	n.d.	1,215	1,215	1,213	1,086	1,010	1,080	820	860	900	n.d.
Percent of world coho production	18	30	74	50	65	72	75	70	73	70	55	46

Note: n.d. = no data available.

3. *Demarcated Fishing Rights*. These authorize the use of a specific area for aquaculture and are critical laws for salmon farming. The applicant must show that the holders of the Common Fishing Rights and Setnet Fishing Rights are satisfied before demarcated rights can be given. Duration of Demarcated Fishing Rights are from 5 to 10 years.

The Fisheries Cooperative Association Law was passed in 1948. That law required fishermen to form cooperatives to receive priority for Common and Setnet Fishery Rights. Once established, the cooperatives became all powerful in allocating Demarcated Fishery Rights for aquaculture purposes to outside fishermen or members of the cooperative. Once obtained, cooperatives must divide demarcated rights among members (usually for a fee) or may conduct aquaculture ventures on behalf of the cooperative; otherwise, the government will lease the rights to individuals or companies outside the cooperative. Cooperatives administer these rights as income producing property with profits divided among members.

Payments to cooperatives derived by fees or leases from farmers are considered compensation for lost fishing areas, but in reality, the revenue generated by the more profitable aquaculture operations often exceeds the foregone fishing income. Cooperatives cannot discriminate on the basis of nationality, sex, or religion when allocating sites, but are allowed to evaluate applicants on factors such as length of residency in the cooperative area and whether an individual has experience in sea farming. From a practical viewpoint, a typical applicant might have to live and fish in a village for 5 years, have been a voting member of the cooperative for 3 years, and have 3 years of culturing experience to have a reasonable chance to receive culture rights.

The Shizugawa Coho Salmon Growers Association typifies cooperatives that feel it is important that the use of ''preferred'' aquaculture growing sites within the sea area of the cooperative do not result in large income differentials. Therefore, demarcated areas are periodically redistributed by lottery and individual farmers are allowed to maintain only three cages.[463] Shizugawa Bay is limited to 90 farms, each with three cages. Similarly, the 123 pen-farmers of the Onagawa Fisheries Cooperative in Miyagi Prefecture are allowed to farm only three cages, but the cages are more than twice the volume of those in Shizugawa Bay.

Several large fishing corporations have shown a keen interest in the profitable coho farming industry and have contributed technical advice to the farmers. But large fishing companies and other large corporations are neither allowed to participate in the coastal fisheries nor be directly involved in farming the coastal bays unless specifically authorized by the fishermen cooperatives. Yet the operation of the salmon farming industry by diverse groups of individual farmers, fish farming cooperatives, and large corporations has functioned remarkably well within the framework of the Fishing Rights System.

The private companies can operate within the framework of the industry through selling eggs and feeds and as fish brokers, and commonly remain involved throughout the life cycle of the farm-reared fish. For example, the Nichiro Company has been importing coho and chinook salmon eggs from Oregon and Washington since 1971. They purchase the eggs and distribute them to the freshwater farmers. Marine fish farmers having farming rights to the coastal bays may buy the smolts from the freshwater farmers through intermediation by the private company.[459] When the fish are harvested the companies may also act as brokers by either directly purchasing the product or taking the product on consignment. Several large companies including Nichiro, Taiyo, and Nichimo contracted with about 350 farmers in 1986 for their production of coho salmon.

IV. LOCATION OF SALTWATER FARMS

Most Japanese coho salmon farms are located along 120 km of the northeast coast of Honshu Island in the Sanriki District which features numerous sheltered bays, natural harbors, and inland seas where mariculture operations are safe from storms — primarily in Iwate and Miyagi Prefectures (Figure 32). Shizugawa Bay in Miyagi Prefecture was the birthplace of commercial coho farming and remains the primary production location in Japan, although several bays to the south, notably Onagawa Bay, also produce farmed coho salmon. Those two bays, plus minor producers like Ogatsu, Samenoura, and Utatsu Bays, also in Miyagi Prefecture, produced more than 85% of the coho salmon farmed in Japan in 1987. Increase in production has been conspicuous in southern Miyagi Prefecture, including Onagawa, Oshika, and Ayukawa, and total production of farmed salmon in that prefecture alone exceeded 16,000 tons in 1989 (Table 12). Southern Iwate Prefecture is the next largest producer.

In recent years other prefectures have attempted coho salmon cage farming. On the Japan Sea, Yamagata, Niigata, Toyama, Ishikawa, Fukui, Tottori, and Shimane Prefectures have produced small quantities of coho salmon in both warmer and colder marine environments than are found in the Sanriki district (Figure 32). Surprisingly, Mie Prefecture in southeastern Honshu has also produced small quantities of coho salmon.

According to Endo,[459] coho salmon farming has succeeded in the Sanriki District for the following reasons:

1. The Sanriki District is an ideal environment for salmon culture, with abundant sheltered bays and inlets. The bays are typical drowned river valleys (rias), sloping rapidly to the open sea, thereby promoting good circulation. Water temperature ranges from 5 to 20°C for most of the year. Although temperatures from July to October are too high for good growth and survival of coho salmon, farmers are able to produce 2.5 kg-fish for local markets in 9 to 10 months of saltwater growout and still avoid high summer temperatures.
2. Fish can be grown to harvestable size and supplied to the domestic markets before wild-caught salmon become available.
3. Farmers and fishermen were seeking new employment opportunities. Prior to 1975, farmers along the coast were engaged in farming oysters and seaweeds such as *Undaria* and *Porphyra,* but because of limits in market demands, were unable to expand. In 1973, many Sanriki District fishermen who had served as crew members on high seas fishing vessels lost their jobs with the imposition of 200-mile exclusive economic zones.
4. Food organisms such as mackerel, sardine, and krill for coho salmon diets were locally abundant and could be supplied at low cost.
5. The larger companies, like Nichiro, were willing to support local farmers in dealing with technical, marketing, and financial problems associated with coho salmon farming.

V. LOCATION OF FRESHWATER FARMS

The total number of Japanese coho salmon smolt farms is reportedly between 200 and 300, but their exact numbers and locations are unknown because so many are small-scale operators who produce coho salmon in addition to traditional species (including masu salmon, "amago" (*O. rhodorus*), charr (*Salvelinus pluvius*), brook trout (*S. fontinalis*), and rainbow trout).

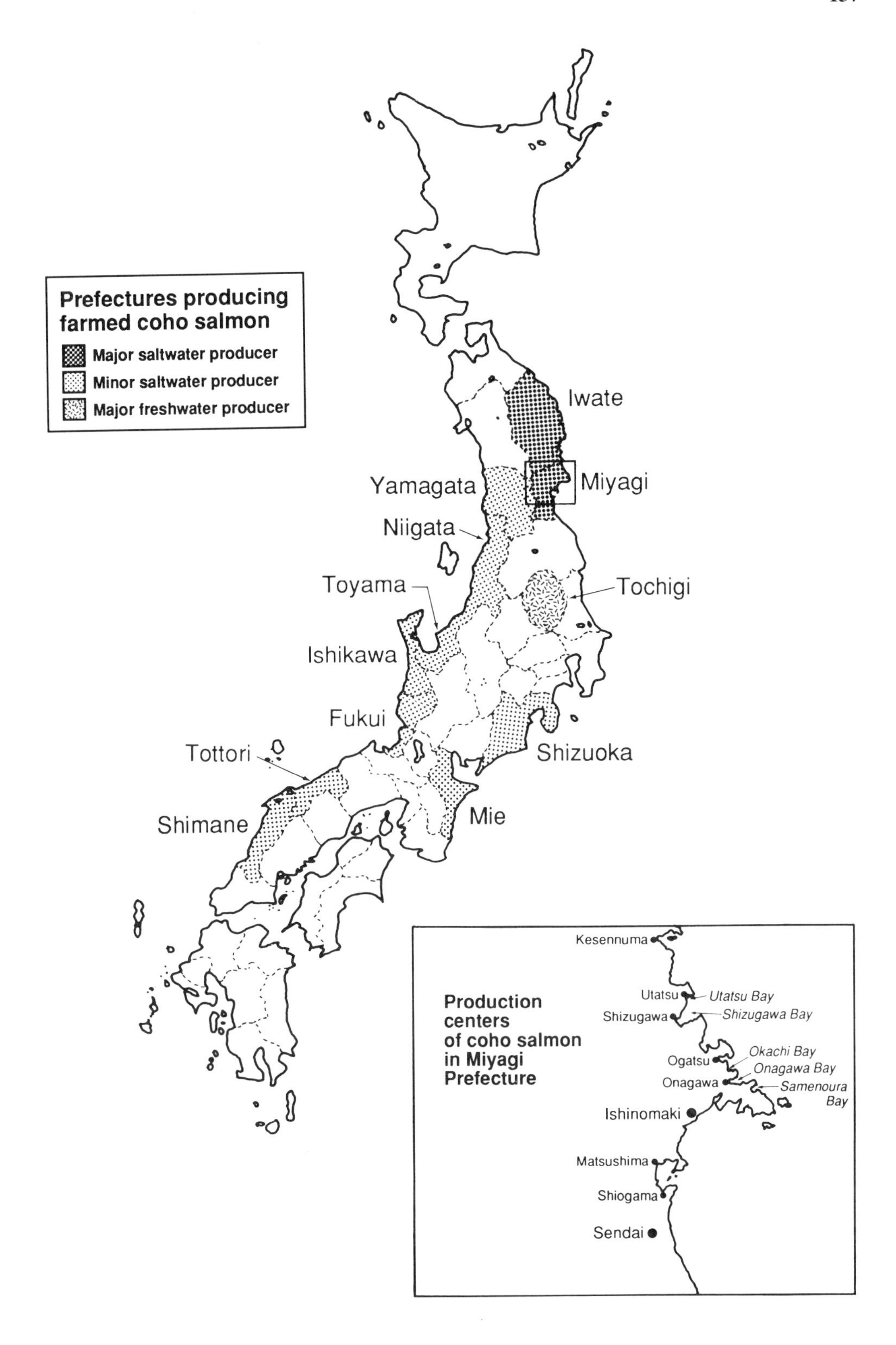

FIGURE 32. Japanese prefectures where coho salmon are farmed.

Most of the smolts are produced in freshwater hatcheries in northern Tochigi Prefecture (Figure 32), 230 km south of Shizugawa Bay. By 1987, the 20 smolt producers of the Tochigi Inland Water Fish Aquaculture Cooperative were producing about 250 tons of smolts per year.[463] Individual farmers produce from 1 to 30 tons of smolts each year depending on the capacity of their facility. Smolts are produced in areas around the towns of Ohtawara, Nasu, and Kuroiso in an area famous for its geothermally heated water and spas. The area was formerly a production center for rainbow trout but many farmers have now supplemented or replaced trout by producing coho salmon smolts. Tochigi Prefecture is a preferred area for growing smolts because abundant groundwater is available with ideal year-round water temperatures of 10 to 15°C capable of producing the required 150-g underyearling smolts between the time of hatch in December and transfer to seawater in October. Some smolts are also produced near Hanaizumi in Iwate Prefecture and Zao in Miyagi Prefecture.

All of the imported eggs for production are from the Pacific Northwest states of Oregon and Washington. The eggs are derived from excess sea-run adults from public hatcheries or sea ranchers. In 1986, 28 million eggs were exported to Japan. Most went to smolt growers in Tochigi Prefecture, but 6 million eggs were reared in Miyagi Prefecture and another 6 million in Iwate Prefecture.[464] By 1989, egg imports exceeded 120 million.

VI. SEAWATER TEMPERATURES AT COHO FARMS IN THE SANRIKI DISTRICT

In Japan, coho salmon are grown outside their native range and are exposed to temperatures much higher than those normally encountered by the species.[458] The oceanography of the coastal seas surrounding Japan is heavily influenced by two major current systems: the warm Pacific Ocean current, the Kuroshio, with its branch, the Tsushima, flowing into the Japan Sea; and the cold northern current, the Oyashio, which flows south out of the Sea of Okhotsk (Figure 33). Oceanographic fronts at the confluence of the northward-moving Kuroshio and southward-flowing Oyashio move latitudinally with the season, and sea surface temperatures in coastal waters rise and fall rapidly in the spring and fall. Japanese coastal waters are typical of western boundaries of northern oceans, with rapid seasonal water temperature changes as well as more extreme highs and lows than are found on the eastern boundaries of northern oceans. The Sanriki District of northeastern Honshu has an annual sea surface temperature range from about 9 to 22°C and is typified by the Sendai temperature curve (Figure 34). By comparison, seawater temperatures in the Oyashio current along northern Hokkaido are too cold for good growth and may approach the lower lethal limit for the species in winter months.

Temperature for safely maintaining greater than 80% of maximum specific growth rate for coho salmon is approximately 12 to 15°C. A growth model for juvenile coho salmon ranging in size from 2 to 30 g predicts maximum specific growth rate at about 15°C.[465] Growth declines above that temperature and approaches zero before the upper lethal limit is reached at 25°C.[466] In Shizugawa Bay, where many Japanese farms are located, 80% of maximum specific growth can be maintained for only about a month of the 9-month growing season, and sea surface temperatures may exceed the upper lethal limit for coho salmon for a few months of each year (Figure 35). These extreme thermal conditions differ in comparison to more moderate western North America farming conditions where summer water temperatures usually do not exceed 15°C (Figure 35).

Despite thermal conditions not entirely suitable for Pacific salmon, Japanese farmers have been able to adapt to local conditions. They have developed a farming system that produces remarkably rapid growth with production strategies for seawater growout in the Sanriki District that avoid high summertime water temperatures. Sea surface temperatures

FIGURE 33. Major Japanese ocean currents.

usually exceed the growth rate optimum by June in Shizugawa Bay. Dissolved oxygen may drop, stressing fish and producing disease; and temperatures may exceed the upper lethal limit by August. Fish can remain at a depth of 5 to 10 m in the cages to avoid the highest temperatures that occur in the surface water layer (0.5 to 1.0 m). Nevertheless, harvest time is dictated by rapidly rising sea surface temperature. Furthermore, the farmers have an economic incentive to harvest farmed fish early in the spring-summer season before wild-caught Pacific salmon drive down the price of farmed fish.

To attain marketable fish weight of 2.5 kg by June, the farmers must transfer large smolts to seawater as soon as seawater temperatures drop below 18°C in late September or early October. Large (150 g) underyearling smolts can be produced after only about 10 months of accelerated freshwater growth. Most farmers prefer 120- to 200-g smolts but will accept smolts as small as 80 g which can hypo-osmoregulate but may not attain harvestable size by June. The saltwater growout system is patterned after that developed in the 1960s in Japan for the culture of rainbow trout in seawater where juveniles were transported to

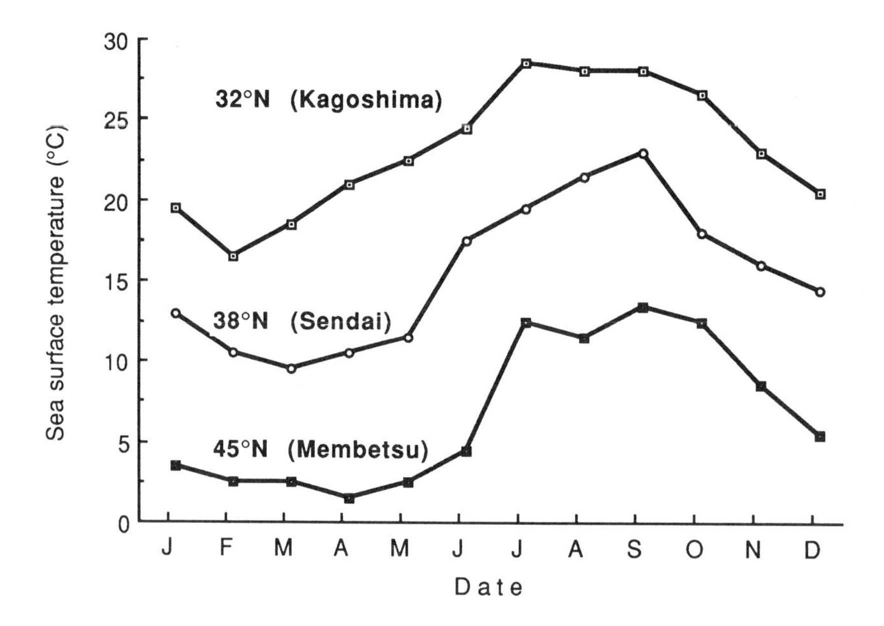

FIGURE 34. Mean monthly sea surface temperatures for selected latitudes on the east coast of Japan (1980).

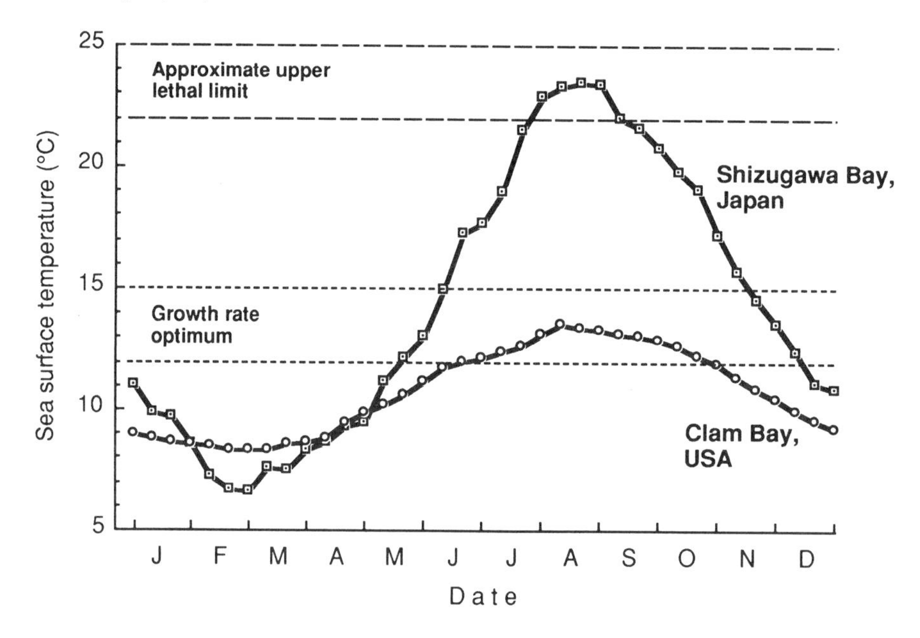

FIGURE 35. Comparison of mean sea surface temperatures in Shizugawa Bay, Mie Prefecture, and Clam Bay, Washington.

the bays and gradually become acclimated to seawater. Rapid growth occurs and first harvesting begins in May after about 8 months in sea cages. Harvest is completed by no later than mid-July when water temperatures once again exceed 18°C.

VII. HUSBANDRY

A. FRESHWATER CULTURE

The culture cycle begins in the states of Washington and Oregon, where eggs are stripped and fertilized at public and private sea ranching facilities and development is accelerated using warm (10 to 12°C) groundwater. Eyed eggs are received in Japan about December and are immediately dipped in 50 to 100 ppm iodophor disinfectant solution in an attempt to prevent transmission of bacterial kidney disease (BKD). BKD is of great concern to Japanese farmers and health certification protocols require North American egg suppliers to inspect for BKD using fluorescent antibody techniques and to destroy infected eggs.

The Japanese Ministry of Fisheries and private companies coordinate egg importations. Eggs are inspected by the Japan Fishery Resource Conservation Association and are quarantined after arrival in Japan. Eggs are imported by some 20 companies — dominated by the largest corporations acting as middlemen — and sold to small private growers and fishermen cooperatives.

Accelerated incubation and rearing of underyearling smolts in Japan is conducted from December of the first year of life to October or November when fish are transported to sea cages. Incubation is completed at 10 to 12°C, and eggs normally hatch in January or early February, depending on incubation temperature. Rearing temperatures are increased to 14 to 15°C at that time. Most smolts are produced in Tochigi Prefecture because the abundant warm groundwater of 10 to 15°C can produce 150-g average weight smolts in the required 10 months.

Rearing containers include rectangular raceways, elliptical ponds, and concrete or plastic circular ponds. Some rearing containers are covered with plastic greenhouses to capture and retain heat.

A variety of freshwater rearing strategies and techniques are used in Japan. Those practiced at the Sanei Hatchery Cooperative on the Mizushiri River near Shizugawa Bay are typical of Miyagi and Iwate Prefectures. The hatchery operates on groundwater but also has access to river water. In 1987, accelerated eyed eggs were received in December, hatched in February, and the fry began feeding in March. Survival to hatch was around 90%; however, with increasing imports and lowered egg quality, survival has recently dropped to around 70%. Fish are hatched in 12°C groundwater and are eventually transferred to circular ponds. Each 300-m^3 circular pond receives about 0.04 m^3/s of water and is stocked with 30,000 fry in February. The fish are switched to river water as fish weight increases and when demand exceeds available groundwater supply.

The need to achieve maximum growth during the freshwater cycle is so finely tuned that the slightest decrease in growth rate due to disease or poor nutrition will result in failure to produce smolts of the required minimum size of 120 g by October or November. Even with accelerated growth, freshwater farmers must grade their fish frequently to attain the average target size of 150 g (120 to 200 g range). By the end of June, most fish exceed 15 g and those below 15 g (usually about 20% of the population) are graded out because they cannot attain sufficient size for transfer to seawater. In September, the Sanei Hatchery eliminates all fish below 120 g (usually 40% or more) since they cannot attain market size by summer; the smaller culls are either harvested for human consumption or released into the Mizushiri River. Most hatcheries in Tochigi Prefecture with warm water grade fish in May, and fish below 15 g are sent to fish meal plants. More than 50% of the initial population can be graded away during the freshwater production cycle. Ponds stocked initially with 30,000 fry may end up with only 15,000 to 20,000 smolts following grading and normal mortality. Iwata and Clarke[467] reported that some Japanese hatcheries successfully rear more than 80% of their stock to 120 g and that some hatcheries retain undersize fish for transfer to seawater the following year.

Feeding is discontinued for 2 days prior to smolt transport to growout bays where they are gradually acclimated to seawater. Some farmers transport fish in 30 to 50% seawater to minimize stress.[467]

B. SALTWATER CULTURE

The transfer of large smolts to seawater in the fall avoids unfavorable summer temperatures in the coastal bays. But the Japanese strategy in producing very large (150 g) underyearlings serves two other purposes as well. Researchers have determined that fish below about 120 g average size may successfully hypo-osmoregulate but cannot attain market size by the time seawater temperatures reach upper lethal limits in July.

To ensure sustained high growth rates, and to reduce osmoregulatory stress, fish are gradually acclimated to seawater. The use of seawater acclimation in Japan is the natural outgrowth of the acclimation that was required to successfully transfer rainbow trout to seawater in the 1960s. It was demonstrated by North American researchers[168] that underyearling coho can be successfully transferred to seawater in spring (May) if they exceed 23 g in body weight. However, they must exceed 50 g in body weight if transferred in the fall (October to November). French researchers, working under environmental conditions similar to those in Japan, have demonstrated that transfer of small coho smolts to seawater of high salinity and temperature (35 ppt, 17 to 18°C) in June or July results in unacceptably high losses, but that fish larger than 80 g were able to survive and grow when transferred in October.[468] Experiments conducted by Japanese researchers confirmed the need for larger fish in the fall,[469] and the Japanese have been able to successfully reduce stress and enhance survival by gradually acclimating smolts to water of increasing salinity.

1. Acclimation to Salt Water

Transfer size averages 150 g, but is adjusted by desired weight of fish at final harvest. Smolts are transferred to acclimation tanks when saltwater temperatures decline to 18°C, usually in October or November. Both land-based and floating acclimation tanks (5 m × 5 m × 2.5 m deep) are used (Figure 36). Each acclimation tank is usually loaded with 10,000 smolts (1.5 ton), which is enough fish for one net-pen (13 × 13 × 10 m deep). Water supply is static, but submersible pumps with flows greater than 0.006 m^3/s provide the required water movement. Tank water is supplemented with bottled oxygen to provide adequate dissolved oxygen levels. During acclimation, fish are held at lower densities than the initial densities in net-pens (1 kg/m^3) and acclimation losses average less than 5%.

Smolts remain in the acclimation tanks from 4 to 6 days. Acclimation to salt water commences after recovery from transportation fatigue. Salinity is usually increased to 40% seawater (approximately isotonic) on the first day and increased 10 to 20% per day thereafter using bay water until full salinity (33 to 34‰) is obtained. When acclimation is complete, the fish are transferred to net-pens via boat from shoreside tanks or towed in the floating acclimation tanks to the cages. Fish are not fed during acclimation, but those acclimated to seawater will begin feeding immediately after transfer to net-pens in the bays.

2. Pen Design and Maintenance

Two major types of net-pens, square and hexagonal, are employed. Square net-pens vary in size between bays; size is determined by the fishermen's cooperatives. For example, in Shizugawa Bay the Shizugawa Corporation (90 families) has individual square pens limited to a size of 10 × 10 × 7 m deep (2,100 m^3),[463] while the Togura Corporation (10 families)

FIGURE 36. Floating seawater acclimation tanks, Shizugawa Bay, Miyagi Prefecture.

uses 30 × 30 × 10-m-deep net-pens.[464] In Onagawa Bay, pens can only be 13 × 13 × 10 m deep. Almost all farmers also grow oysters, scallops, tunicates, and seaweeds.

Net-pens are commonly constructed with frameworks of 5- to 7-cm diameter steel pipes with flotation provided by styrofoam blocks (Figure 37). Mesh size is 2 to 3 cm (stretched). Single pens are common. At maximum, usually no more than three pens are hooked in tandem with adequate (10 m) spacing between them. Japanese net designs are different from those used in North America. The nets are hung on the square and the net bottoms are sloped so mortalities collect in the center. Mortality removal is facilitated by use of a net-covered hoop (80 cm diameter) rigged with a bridle for collection and retrieval. Net rotations, substituting clean nets for biologically fouled ones, are performed as soon as fouling is observed. During the spring and summer, rotation may be performed as frequently as twice monthly. Cages are covered with netting to keep out predatory birds.

3. Loading Densities

Square cages, 10 × 10 × 7 m deep or 13 × 13 × 10 m deep, are initially stocked with 5,000 and 10,000 150-g smolts, respectively. These initial stockings conform to recommended densities of about 1 kg of fish per cubic meter. In reality, farmers often stock at higher densities, but 8,000 150-g smolts in 10 × 10 × 7-m-deep cages (7.7 kg/m^3) is considered too high. A density of 10 to 15 kg/m^3 at harvest is recommended.

4. Fish Growth and Survival

The Japanese system of salmon culture depends on rapid sustained seawater growth to marketable size. Growth stasis due to marine diseases, low or high environmental temperatures, poor physiological condition of smolts, or self-pollution will result in late harvest of undersize fish. Fish undergo rapid growth during their short tenure in saltwater (Figure 38).

FIGURE 37. Sea cage, Shizugawa Bay, Miyagi Prefecture.

First harvest usually occurs after 6 months of rearing in May with 2.0-kg fish (Figure 39) and ends early in August with 3- to 3.5-kg fish. Length of saltwater residence is determined by temperature. Fish are harvested before water temperatures reach 20°C, usually in July (Table 13, Figure 40).

The mean survival rates are reportedly 90% to swim-up, 80% from swim-up to smolt, and 80% from smolt to harvest. Greater than 90% survival and less than 55% survival from smolt to harvest have been reported. Vibrio, furunculosis, and BKD are the most frequently encountered diseases during saltwater culture. BKD has been identified as the most serious saltwater disease and is most evident as temperatures increase in the spring and summer. In recent years, mortalities as high as 30% during saltwater culture have been blamed on BKD or BKD-exacerbated diseases. Farmers blame U.S. egg sellers for the rising incidence of BKD in farmed coho. Whether a function of better detection procedures or not, the consensus appears to be that BKD is a relatively recent issue. Furunculosis has also been responsible for some initial losses after transfer. Vibriosis, *Costia,* and gill amebae were not considered as contributing significantly to seawater mortalities. A nutritionally related disease (''fatty liver disease''), which leads to anemia, is sometimes mentioned as a problem.[464]

5. Diet and Rations

Moist pellets manufactured by local farmers are used for feed in coho salmon culture in Miyagi Prefecture. The feed is compounded from a commercial dry diet and raw fish. Researchers have found that the best feed efficiency is obtained with a 50:50 mix of raw fish and dry diet. However, fish farmers generally use a mix containing only 10% dry diet to economize.[464] The use of a high percentage of raw fish in the diet results in a large drip loss and can create local ''self-pollution'' problems in the bays. Feed is made fresh daily with pellet size determined as 70% of the maximum gape of the average fish's mouth.

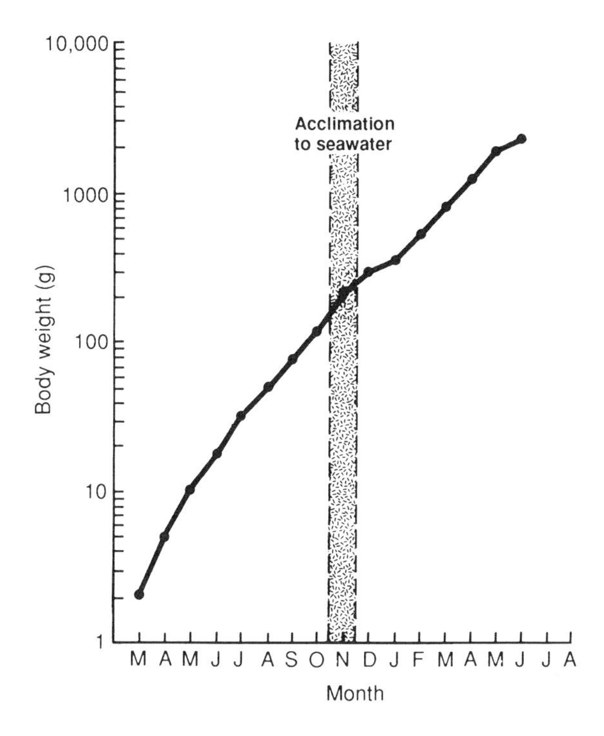

FIGURE 38. Typical freshwater and saltwater growth of farmed coho salmon in Iwate and Miyagi Prefectures (modified from Iwata and Clarke[467]).

FIGURE 39. Small market-size coho salmon raised in sea cages in Shizugawa Bay.

State-of-the-art feed processing machines are commonly used. These may be land-based, but increasingly are installed on the feeder boats at costs of $20,000 to $30,000. Raw constituents are ground aboard ship, mixed with bagged dry constituents, and extruded into containers on the deck after which they are fed directly to the fish.

The raw fish constituent is composed largely of frozen mackerel and sardines, but may

TABLE 13
Growth, Feed Conversion, and Survival of Coho Salmon Growth at three Farms in Shizugawa Bay (October 1981 to July 1982)[471]

	Mean monthly temp (°C)			Av fish weight(g)			Approx feed conversion (wet to wet)			Cumulative % mortality		
Farm	**1**	**2**	**3**	**1**	**2**	**3**	**1**	**2**	**3**	**1**	**2**	**3**
Oct	16.1	—	—	167	—	—	—	—	—	stocking	stocking	—
Nov	13.4	12.3	12.4	199	185	152	5.7	7.5	—	stocking	0	stocking
Dec	10.0	10.0	10.0	276	246	199	3.8	4.7	4.2	0	1	stocking
Jan	7.9	8.0	8.0	386	352	282	2.9	4.2	3.3	1	1	1
Feb	6.5	6.6	6.7	526	494	392	7.2	8.5	7.9	1	1	1
Mar	6.5	6.6	6.6	725	694	550	3.4	3.4	3.9	2	1	2
Apr	8.1	7.9	7.7	1005	986	776	2.6	3.7	2.8	4	2	2
May	11.2	10.9	10.3	1393	1400		—	—	—	12	2	3
	1096											
Jun	16.0	15.7	15.5	1930	1988		—	—	2.5	51[a]	24[a]	6
	1547											
Jul	19.5	18.0	19.4	2503	2675		—	—	—	99[a]	72[a]	39[a]
	2120											

[a] Harvesting

FIGURE 40. Harvest of farmed coho salmon, Shizugawa Bay.

also contain saury and juvenile cotids. The proportion of fish species in the feed will vary from farm to farm and month to month throughout the growing season depending on availability (Figure 41). For marketability, red flesh is necessary when cultured salmon are harvested. Flesh color is examined in March and crustaceans containing the pigment astaxanthin are added to the finishing diet and may comprise 10 to 20% of the feed from March until harvest.[470]

According to Toole,[464] a typical dry mix is composed of 45% fish meal, 28% soybean meal, 14% rice, and vitamins. Vitamin B_1 is added to the dry mix to compensate for the loss of vitamins due to thiaminase activity in the sardine component of the feed. In addition, the dry mix may contain folic acid, synthetic pigments, 2% charcoal (presumably to decrease ''fishy'' odor and absorb possible pollutants before they fall to the sea bed beneath the pens), and 2% brewers yeast.

Fish are fed once per day from November to April when water temperatures are lower than 10°C and twice daily from May to August when temperatures are above 10°C. A feeding rate of 2% body weight per day is recommended during colder months and 3 to 4% during warmer months. Feed conversions of 2.5:1 to 8.5:1 with a mean of about 6:1 have been reported[471] (Table 13), but feed conversion is less important economically than the need to attain the earliest marketable size. Consequently, farmers may feed to satiation at rates of 5% of body weight per day and above during most warmer months.

6. Brood Stock

The majority of eggs are imported. Researchers feel this trend will continue as long as egg prices remain the same for imported and domestic eggs, but recent concerns over diseases imported with the eggs may accelerate brood stock development. A limited number of eggs have been taken from brood stock (300,000 in 1985 compared with imports of approximately

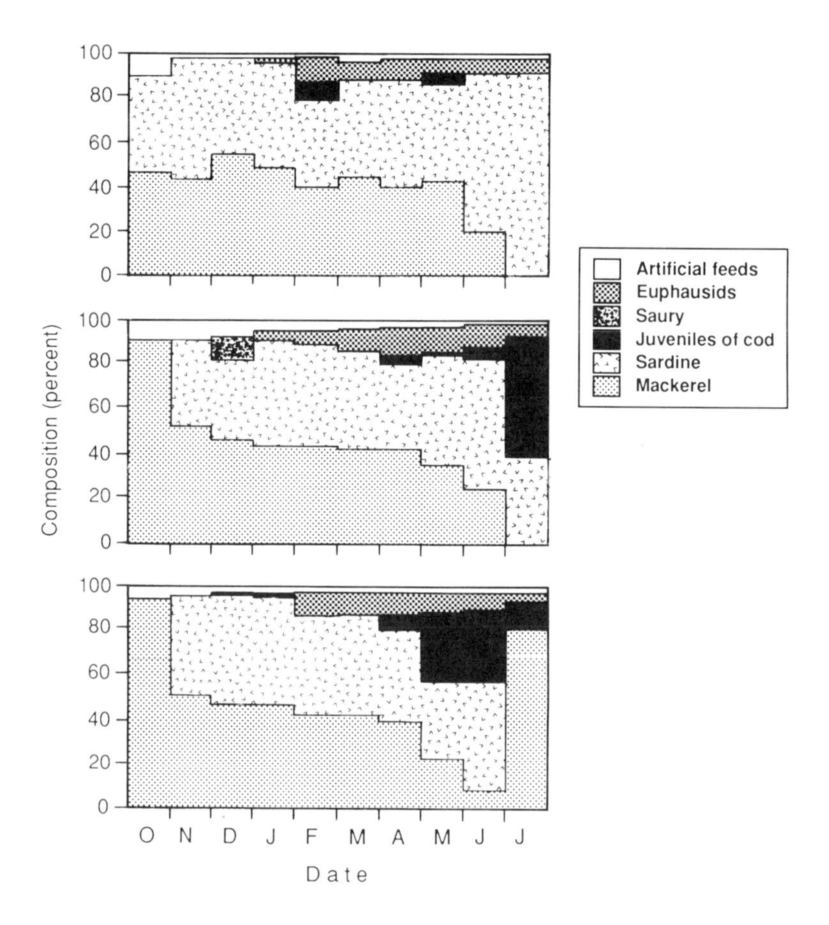

FIGURE 41. Seasonal changes in feed composition at three saltwater farms, Miyagi Prefecture.[471]

18×10^6). The maturation and spawning procedure is to sort out the largest production fish in June, July, and August, and transport them in 50% seawater to freshwater farms for final maturation. Typically, coho brood stock are held in spring water with a pH of 7.0 and a temperature of 12°C.

7. Self-Pollution

The bays in Miyagi and Iwate Prefectures are selected for coho farming on the basis of their good circulation with the open ocean, but in recent years, as farming activities have expanded, the bays have become overloaded with coho cages and shellfish rafts. As a result, water exchange in the inner bays has been reduced.

Organic material from uneaten feed and fecal wastes has accumulated in some bays, resulting in anoxic condition of the sediments directly below the cages. The problem is monitored by assaying for H_2S at the seafloor. Some outgassing of H_2S from the seafloor has been observed. Dissolved oxygen is also monitored near the seafloor-water interface, but the best indication of increasing levels of H_2S and lowering of dissolved oxygen is given by survival of oysters and scallops cultured on vertical ropes in the bays. When dissolved oxygen begins to drop, usually near the seafloor, scallops and oysters on strings near the sediment-water interface begin to die and thereby act as a good (although unwanted) natural bioassay. As the coho farmers are often also the scallop and oyster farmers, conflicts are minimized, but the problem persists. Attempts have been made to minimize self-pollution by moving coho cages from inner to outer bay regions, but some price is paid by the farmers because the outer bays are less protected and cooler, so fish growth is slightly reduced. Some pens have been permanently moved to deeper water to alleviate self-pollution problems associated with the intensive finfish and mollusc aquaculture activities in Shizugawa Bay. A recommendation has been made that in bays with poor flushing (1.3 m tidal range and less), pens should be spaced a minimum of 40 m apart.

Chapter 7

POLICY ISSUES WITH RESPECT TO SALMONID CULTURE

David Fluharty

TABLE OF CONTENTS

I. INTRODUCTION

The development of government policies pertaining to the culture of salmon and trout has evolved within the context of management policies for all species of fish and the culture of aquatic organisms. This reflects the enormous value that salmonid fisheries have around the world as a preferred consumption good, a source of income, a recreational resource, and, increasingly, a symbol of environmental quality. In each of the major geographic areas where salmon and trout exist as wild stocks, there is active culture activity. In fact, in many areas the role of culture in the form of hatchery support for recruitment is so commonplace that it is taken for granted. Extension of Pacific salmon culture, e.g., to the Great Lakes of the United States and coasts of Chile and New Zealand, has allowed development in areas where salmon are not indigenous. Similarly, Atlantic salmon are increasingly the culture species of choice throughout the world because of the apparent relative ease of cultivation.

At this juncture, the future of salmonid culture is quite strong as measured by market demand and societal interests, but there is an increasing array of policy issues being raised that call for analysis, communication, and innovative response for that future to be realized. The biological, technological, and managerial components of salmon and trout culture, while still improving, are not the major problems facing this activity. Positive technical change in one policy arena may result in problems of adjustment in others. In addition, mistakes and mismanagement may cause economic and environmental problems. Social, political, and economic constraints are increasingly important. However, there is relatively little information gathered on a systematic basis on the development of policies for salmonid culture. Most of the literature on this topic is scattered in government reports and proceedings of conferences. Monographs by Bowden,[472] Wildsmith,[473] and Centre National pour l'Exploitation des Océans[474] provide comprehensive, but somewhat out-dated treatment of aquaculture law and policy in the United States, Canada, and France, respectively. White[475] reviewed a number of issues of relevance to salmon culture on the west coast of North America.

Salmon culture policy can be characterized as having two major purposes. First is to promote the development of salmonid culture through provision of conditions necessary to carry out operations. This may include securing property rights in areas for development; establishing research, veterinary services, and other infrastructure to support operations; and providing subsidies through fiscal arrangements to achieve other societal objectives. The second purpose of policy is to ensure that conflicts with other uses and activities are minimized. This can be seen in the case of regulations on siting, waste discharge, etc. In addition, it should be noted that salmonid culture is benefited and impacted by the existing legal matrix developed for operating any sort of business (insurance, banking, taxation, patents, criminal law, etc.).[473] However, unless this latter category of policies has a particular impact on culture of salmonids that distinguishes it from others, it is not treated herein.

A variety of policies are neither legislated nor made part of regulations of public agencies but come from recommendations of professional societies and the literature on "best management practice". These can be very effective in informally avoiding or dealing with problems outside of the public policy process. See, for example, the American Fisheries Society position statement on Commercial Aquaculture.[476]

At the outset it is acknowledged that stated policies are not always implemented, enforced, or in other ways put into effect. This can be a result of lack of government will, scarcity of human and financial resources, and other factors. No attempt is made to be comprehensive in describing national policies and approaches and only limited analysis is pursued.

The evolution of policies for trout and salmon culture has a long history. However, this chapter gives prime focus to the policy questions surrounding modern culture of these species. Treatment is by generic topics which explore the issues and approaches. Examples are provided as illustration.

II. HISTORICAL DEVELOPMENT

Salmonid culture policy is dealt with only briefly here as background for the main focus on contemporary policy issues. There appear to be roughly three periods in the development of salmonid culture — prior to the 1940s, the post-World War II period, and what might be termed the modern period starting in the 1960s. The dates are somewhat arbitrary but in terms of policies, there is some utility in this division.

Prior to the 1940s, efforts to culture fish were relatively low on the learning curve but highly popular. Declines in inland fisheries and market opportunities prompted efforts to develop salmonid culture to augment natural production beginning in Europe in the early 1800s and spread to North America in the mid-19th century (see Chapter 5 and Bowen[477]). Restoration of fisheries through stocking of fish had considerable popular appeal as opposed to the institution of regulations which were difficult to enforce and did not produce immediate results. During this early period there were virtually no policies with regard to culture activities in Europe or North America — eggs and fish could be transferred at will. Through the activities of state fish commissions and the newly established U.S. Fish Commission, public hatcheries were developed in the 1870s, and the culture of salmonid stocks began. Salmon eggs collected in northern California were placed in virtually all major river and lake systems with the expectation that they would produce immediate results. Coincidentally, fish culturists on the east coast noted the first occurrence of fish diseases (furunculosis) in brown trout just following the introduction of rainbow trout from the west coast.

Results actually obtained from the hatching and stocking of salmonids were disappointing. By the 1930s it is generally conceded that both in Europe and North America those efforts failed because eggs and fish were planted under conditions where survival was extremely low.[478] The hatchery program in British Columbia was actually stopped in 1935 because of doubt about its effectiveness.[478] Despite the sometimes negative results of these experiments carried out with public funding, commercial trout farming did bring modest success to some operators in the vicinity of urban markets. Public and scientific support for government involvement in fish culture remained strong.

Hatchery work continued with research on diets, disease control, and culture techniques through the 1940s and 1950s. In addition, the expansion of major hydroelectric projects on rivers in Scandinavia and North America prompted considerable research on ways to circumvent the impacts of the dams. By the 1960s those efforts resulted in the development of new feed formulas and permitted the improvement of salmon hatchery practices in places like the Columbia River by lowering the cost of feeding fish. Thus, it was possible to raise fish to a larger size before release, which increased survival rates. The techniques developed in the 1940s and 1950s were also a boon to development of commercial rainbow trout culture in the northwestern United States[480] and led to increased success in recreational trout stocking programs.[481]

Development of rainbow trout sea culture in the late 1950s and experimentation with the culture of Atlantic salmon in the 1960s by private operators led to technological breakthroughs.[353] In the United States, public policies supported work to develop improved ways to raise salmon and resulted in the establishment of commercial net-pen culture in Puget Sound.[482] From these beginnings there has been particularly rapid expansion of salmonid culture worldwide that has been either aided by government policies or hampered by them.

In the remainder of this chapter, the evolution of the major policy questions raised in the development of modern fish culture is addressed.

III. GENERAL CONSIDERATIONS

Public policies set the direction for salmonid culture, usually balancing between efforts not to impose unnecessary restrictions on the development while at the same time protecting wild salmonid stocks and other public uses. Because of the vastly different national contexts that exist, national rules may vary considerably depending on legal, political, and institutional arrangements as well as characteristics of the environments, societies, and economic systems. It is useful to examine two important aspects of salmonid culture — property rights and governmental jurisdictions at the outset.

A. PROPERTY RIGHTS

Property rights are one of the first things to be determined when considering culture operations.[473] Because fish stocks generally have been considered wild (*ferae naturae*) as opposed to domestic (*domitae naturae*), under common law, ownership is established through capture. Wild and released anadromous salmonids complicate the issue of ownership in ways that have not been fully resolved in many countries. Public investments in hatcheries for restoring the common access fisheries have been justified in support of commercial capture fisheries and recreational fishing. However, without some proprietary interest protected by law, private parties have little incentive to invest in culture practices. Thus, one of the first policies that must be established is the property right in the stock under culture.

Lack of control over ocean harvests is a major drawback to ocean ranching, for example, that seems unlikely to be resolved in favor of the culturist in North America even with improvements in tagging and other measures. The culturist must be able to control the movements and capture of the fish throughout their life cycles. The situation is further complicated by loss of cultured stock due to *force majeure,* mistakes by the manager, and predation. Any fish loss from culture operations can be the gain of any party capable of capturing it.

Fish culture operations also require space. The arrangements for use of that space are also subject to property rights. For tank or pond culture on land, the fee simple ownership of property establishes the right. This is the chief method used in trout culture. However, arrangements for use of space (surface, water column, and bed) in lakes, streams, and tidal waters, where salmonid culture regularly occurs, can be extremely complicated. Lacking fee simple ownership, some alternative arrangement must be developed that establishes a proprietary connection with the space needed; e.g., lease or license. Acquisitions of such rights are subject to a variety of other bodies of law relating to public rights of navigation and fishing and state ownership of those areas. In addition, a considerable body of law exists with respect to the rights of adjacent foreshore and subaquatic land owners.[473]

Of course, there are considerable differences in the property arrangements for salmonid culture depending on country and state or province. The foregoing discussion is based on common law countries like Canada, the United Kingdom, and the United States. Iceland assigns property rights in rivers to adjacent land owners who must be part of fishing associations that manage the resource.[483] In Japan, salmon in the rivers of Hokkaido are reserved exclusively for use by government fish hatcheries and a few fishing cooperatives.[484] While there appears to be considerable flux in arrangements for salmonid culture in the Soviet Union that may eventually lead to establishment of property rights in fish or culture sites, at present all rights are held by the state.[485-487]

B. JURISDICTION

Perhaps the most arcane area of public policy for management of salmonid aquaculture is determination of which agency or agencies have jurisdiction. Seldom is there a single agency that controls all aspects of salmonid culture. Generally there are split authorities between national and state or provincial level authorities. Probably the most important functional policy questions involve who is allowed to culture salmonids and who controls the property rights in areas owned by the public. Almost always, these are different agencies. Salmonid culture includes the production of fish for commercial and sport fisheries through public hatcheries. This complicates the situation when it comes to regulation of private fish and shellfish farming where the activity is considered the same as producing an agricultural crop. Thus, while the functions that need to be performed may not vary much from country to country, the administrative means devised to accomplish them may vary considerably. A few examples illustrate this point.

Salmonid culture in Scotland is subject to a number of agency responsibilities.[487] As part of the United Kingdom system of government, the Cabinet level Scottish Office has general responsibility exercised through two of its departments, the Scottish Development Department (SDD) and the Department of Agriculture and Fisheries (DAFS). The SDD handles general planning and consults on control of the allocation of seabed sites. DAFS deals with the fish farming industry in Scotland and with other Scottish agencies on fish farming. DAFS administers disease control legislation and performs research and development. It works directly with the European Community's (EC) Commission on projects under the Commission's guidance program for aquaculture. Assignment of leases to salmon culture in internal and territorial sea waters of the U.K. is the role of the Crown Estate Commissioners. Local planning authorities may also require application for a permit to use sites that impinge on the low water mark. In rivers, control over the extraction of waters for fish farming and control of pollution are the functions of River Purification Boards. Economic development programs for salmonid culture may fall under the jurisdiction of the Highlands and Islands Development Board and may be eligible for financing by the Scottish Development Agency which has assisted in development of quality assurance and marketing.

In the United States, Washington State is an example of how jurisdiction is divided among a wide range of federal, state, and local authorities.[304] Federal agencies include: (1) the U.S. Army Corps of Engineers, with responsibility for permits regarding activities affecting navigation (reviewed also by the U.S. Coast Guard) and discharge of materials into navigable waters; (2) the U.S. Fish and Wildlife Service, which consults on Corps permits with respect to protection of fish, shellfish, birds, mammals, and their habitats; (3) the Environmental Protection Agency, which provides oversight over the Washington Department of Ecology administration of the pollution control permit system; (4) the Food and Drug Administration, which ensures quality of farmed fish in interstate commerce and approves chemicals in fish farming operations; and (5) the National Marine Fisheries Service, which reviews the Corps permits with respect to fish and marine mammals. State agencies include (1) the Washington Department of Fisheries with lead control for all marine species, disease control, registration, and statistical control of aquaculture; (2) the Department of Natural Resources, which manages state tidelands and beds of lakes and rivers with the aim of providing public benefits and revenues; (3) the Department of Ecology, which is in charge of water quality control programs; and (4) the Department of Agriculture, which, although it does not issue permits, is responsible for promoting the development of aquaculture and assisting with marketing products. Local governments administer local shoreline master programs and environmental compliance under the state environmental policy act. Local government zoning and building codes may apply to upland portions of aquaculture facilities.

Suffice it to say that salmonid culture policy, especially with respect to salmon net-pens, is a regulatory maze. It is small wonder that there are constant complaints about the difficulty of obtaining all necessary permits and repeated recommendations for some way to simplify or consolidate the application process.

IV. POLICY ISSUES

Evolution of salmonid culture policies has been particularly rapid in the last 20 years as legislative and administrative bodies have been forced to take advantage of opportunities and to solve problems brought on by the development of commercial salmon culture. Nations with major programs like Norway and Canada have obviously devoted considerable resources to policy development. However, size of program is not the only indicator of activity as seen in areas like the states of Alaska and Washington in the United States where development has been highly contested. Faced with the dilemma of describing national approaches to policy making or focusing on policy issues generically, the decision is to do the latter using a composite of national experiences to illustrate the range of approaches. In this section, the intent is not to exclude developments in trout culture; however, there seems to be relatively less change taking place in the policies affecting that portion of salmonid culture. Certain aspects do come into play with respect to new policies for water quality and quantity and possibly with respect to genetic integrity of stocks and approaches to fish stocking in natural environments. The fee-simple ownership of land used for trout culture and the less significant role that cultured trout play in international markets significantly reduce the level of controversy and pace of policy development.

Policy issues for salmonid culture have been divided into nine topical areas. These include licences/leases, navigation conflicts, fisheries conflicts, water quality/quantity, subsidization, genetic questions, control of disease, predation, and recreation/visual quality odor/noise. There is considerable overlap among some of the issue areas; however, it may be useful to maintain an arbitrary distinction for purposes of discussion.

A. LICENSES/LEASES

Generally, it is necessary to have a license to culture salmonids and a lease over an area in which to perform the operation. As noted above, these may have to be obtained from two or more separate agencies. In these processes it is possible for governments to exert policy control over the culturist and the siting of the salmon farming operation.

A license is almost always required for a fish culture operation even if it is on private land. Usually the license requires the culturist to provide a detailed plan of the fish farm and intended production. This ensures that all technical requirements for equipment and design are met. Any significant changes in the design or proposed production usually require amendment of the license. Holders of licenses must agree to certain reporting standards and operation procedures, e.g., stocking densities, disease control, inspection, transfer and transport of organisms and products. Failure to comply with the conditions can result in fines or loss of license. The license term is usually annual to ensure that reporting is timely and adequate.[304] Separate licenses may be required for different aspects of the farming operation.[353] The license can be a particularly important element in the policy regime because it allows government agencies to deny a property owner the right to cultivate fish if the site chosen is inadequate.

A lease or other instrument that grants lawful access is required if the farm location is not privately owned. In most cases, salmon culture operations require deep enough water to be outside the normal property zones and intrude into areas owned by the general public. This gives government agencies a powerful tool with which to ensure that the aquaculture site meets technical design requirements for a culture operation. It also allows a variety of other factors to be taken into account when siting culture operations, such as location relative to other farms, viewshed, interference with navigation and fishing, etc. Some attempts have been made to systematically identify areas that are suitable for aquaculture operations and to direct growth into those areas.[488] Government agencies are not able to be totally in charge of siting, however, because most fish farming applications involve the use of an upland site for staging. Thus, a property owner may want to use waters adjacent to his or her holdings. What results is a negotiation on siting with considerable discretion available to the agency.

The lease is a contract between the government and the culturist. It is a negotiated agreement that specifies the rights and duties of each party. The term of the lease, size of area, fees, assignability, and conditions for renewal or termination are spelled out. Conditions demanded by the government for use of public lands can be relatively generous in the event that it wishes to encourage fish culture or onerous if its goal is to discourage the activity at a particular site. Generally, an attempt is made to collect a fair market value for the site.[473]

B. NAVIGATION CONFLICTS

The very features about a site that make it attractive for fish culture (water depth, circulation, access) may coincide with navigation channels. In the United States, control over the permitting of structures in navigable waters is vested in the U.S. Army Corps of Engineers. The U.S. Coast Guard advises the Corps on the impact on navigation and may require that certain aids to navigation be installed.[472] Fish farms may break loose from anchors and cause a hazard to navigation by floating into the path of ship traffic.[304] Because fish culture facilities are located in protected areas, they may displace vessels from anchorages used during rough weather. Siting criteria for aquaculture facilities generally advise the avoidance of conflict with navigation. There is scant empirical data on interference with navigation by fish culture sites.

C. FISHERIES CONFLICTS

Salmonid culture may conflict with various aspects of commercial and recreational fisheries. Hatchery production, ocean ranching, physical displacement of harvesting activities, and market competition with harvested fish represent the principal areas of concern.

1. Public Hatcheries

Public hatchery policies are seen as a potential benefit to commercial harvests and as a possible threat. On the benefit side, estimates of the percentage of commercial fisheries catch of salmon contributed by hatcheries range from 25% for Atlantic salmon in the Baltic fishery to 80% for chum salmon in Japan[489] and 20 to 30% of all Pacific salmon in the North Pacific.[490] The enormous investments in fish hatchery production to compensate for losses due to hydroelectric power development continue to make large contributions to commercial harvests.

The threat posed by hatchery production comes from the concern that hatchery production may be used to offset losses of wild stocks due to other things like dam construction and habitat destruction. Recognition of the importance of protection of genetic stocks for wild fish (see below) raises concerns that hatcheries will be seen as a cheaper policy alternative for fish production and efforts to protect habitat for natural stocks will stop or be impaired strategically.[491]

2. Ocean Ranching

In this section the term "ocean ranching" is reserved for private efforts to hatch and release salmonids for harvest on their return from the sea. In this case, it differs from a public fish hatchery with respect to the hoped for private profit from the use of the commons.

The approach has been tried with varying degrees of success that seem directly related to whether or not there are extensive high seas or nonterminal area fisheries for the ocean-going stocks. Without alterations in policies that would prohibit all salmon fisheries at sea, legalize trap fisheries, permit harvest of private ranched fish during closed wild stock seasons, and define ownership of privately released stocks, ocean ranching appears to be nonviable.[489,492]

3. Physical Displacement

The siting of culture facilities in areas where fishing takes place or where fishing boats customarily anchor can result in displacement. Similarly, a scattered configuration of net-pen culture could make for difficult maneuvering with fishing gear. Under such circumstances, considerable hazard exists for damage to fishing gear, like driftnets, and the culture facilities. The possible result of this problem could be loss of catch and allocation under certain conditions. Coordination among other users would be a necessary policy approach to dealing with this issue.[304]

4. Economic Displacement

In areas like northwest North America, where there are still large quantities of wild stocks, the issue of competition from farmed salmonids is perceived as very real. This has played an important role in the development of salmonid culture policies in the states of Alaska and Washington.[493,494] Commercial fishing interests are concerned that farm-reared salmon will out-compete wild salmon and that this will result in a long run decline in their industry. Most of the studies done seem to discount that risk.[495] In fact, the failure to develop fish farming under the favorable environmental and market access conditions in Alaska results in a net loss. Additional arguments used against the establishment of fish farming include concerns about communication of disease from farmed stocks to wild stocks and of straying of salmon lost from culture facilities. Again, these are not considered to be insurmountable problems and certainly not ones that are unique to Alaska. In neighboring British Columbia, for example, some off-season fishermen obtain employment in culture operations.

D. WATER QUALITY/QUANTITY

Salmonid culture presents an interesting problem with respect to water quality. Culture operations depend on having high quality water for the pollution sensitive fish being raised. Still, fish wastes and food wastes from culture operations have resulted in pollution problems in the vicinity of poorly sited facilities. Theoretical energy and materials balance analyses show that even the best managed and most efficient culture facilities produce wastes. Empirical studies also show that dissolved oxygen declines and nutrient levels increase under net-pens where there is poor circulation. Onshore trout culture operations in the United States have been required to apply for discharge permits under the Clean Water Act. State agencies and the Environmental Protection Agency are considering a similar requirement for net-pen culture operations if alternate ways cannot be devised to diminish fecal and food waste loading of the bottom sediments.

Other water quality concerns associated with net-pen culture include the spread of antibiotics, hormones, bacteria, and disease organisms into waters surrounding the facilities. Studies of these phenomena seem to indicate that the combination of low level discharges and high dilution rates lead to the result that risks of pollution and human health impacts are negligible.[304,496] Use of toxic chemicals like tributyl tin (TBT) as an antifoulant has been drastically reduced and replaced by mechanical treatment. Proper siting of culture operations and improved feeds and feeding methods have all contributed to a reduction of effects.

Still, questions are being raised about the balance in policies for water quality cleanup of sewage and at the same time permitting activities, like fish culture, that add substantial quantities of the same nutrients. Policy discussions have been raised about the obligation of public agencies to assure water quality in areas where they license or lease fish culture sites.[473] In some cases, the water quality requirements of fish farming are seen as providing impetus for maintaining water quality in areas increasingly affected by nonpoint source discharge.

Of particular concern is the presence of major algal blooms that have had disastrous effects on salmonid culture facilities in the late 1980s in both Scandinavia and the west coast of North America. The origins of these blooms and the relationships to nutrients, temperatures, and other factors have not been explained. But many are concerned that human activities contributing to pollution may have a role[341,497]

Water quantity is a recurring issue in the culture of trout. The need to have a large quantity of high quality water for trout culture is one of the factors that has given regions like the Hagerman Valley of Idaho a comparative advantage. Expansion of production to meet demand will require tapping additional sources of water. In the western United States, good quality surface and subsurface waters are scarce.[480,498]

E. PROVISION OF SUBSIDIES

Salmonid culture has benefitted from considerable government investments in research, infrastructure, advisory and professional service, and financial arrangements. Each of these investments is the result of public policies to develop, promote, and control culture activities paid for through public revenues. Given the often inextricable linkages between private and public activities it is impossible to fully judge the impacts of these apparent subsidies. In Norway,[499] Scotland,[500] and Canada,[501] various forms of subsidy are seen as part of social and economic policy aimed at increasing employment and wages in rural communities. Tribes in Washington and British Columbia see aquaculture as a potential instrument of social and economic policy that is consistent with developmental aspirations.[502]

F. GENETIC QUESTIONS

Recognition that each stock of salmon is selectively adapted to its natal waters has raised questions about hatchery management practice as well as straying of hatchery- and farm-raised fish *vis à vis* wild stocks. These issues have apparently been more important in the public debates in the Pacific Northwest of the United States than in countries like Norway although considerable scientific research has been done in both North America and Europe.[362] Public policy in this area is being pushed and pulled by scientific and popular debate alternating between ''any loss of genetic integrity is a potential disaster for salmon stocks'' to ''genetic manipulation is the solution to production problems of salmon and trout.''[503]

Another issue related to genetics is the introduction of new salmonid species. The early history of massive trout transfers is one example. The rainbow trout native to the western United States is now the basis for the European trout industry. With notable exceptions like the introduction of Pacific salmon in the Great Lakes and Atlantic salmon in Argentina and the Faroe Islands, salmon transfers have had a remarkably poor record. Deliberate attempts to introduce Atlantic salmon into the wild in 36 countries have failed in all but the two named above. There seems to be consensus that escapement and loss of Atlantic salmon from pen culture can result in survival. Losses should be avoided and monitored, but the likelihood of these fish establishing self-sustaining runs is small as long as there is competition from Pacific salmon and native trout species. In addition, there is little probability of Atlantic salmon stocks interbreeding with Pacific stocks as laboratory experiments have demonstrated the species to be genetically incapable of producing viable offspring. The chance that Atlantic salmon could breed with steelhead or cutthroat trout is also small because they appear to breed at different seasons.[304]

G. DISEASE CONTROL

Besides pollution issues, the policy issue that seems of most concern to the general public and to many fisheries professionals is fish disease. This issue has multiple dimensions. Among them are

1. Introduction of fish diseases through transfer of fish and their roe,
2. Concern over transmission of fish diseases or parasites from wild stocks to cultured stocks and *vice versa,*
3. Human health risks, and
4. Inadvertent effects of disease control outside of the culture area; e.g., dispersal of antibiotics.

With respect to the introduction of exotic diseases, many states and countries restrict fish culture operations to indigenous species and prohibit the importation of live fish. Some permit the importation of eggs under strict conditions of inspection and documentation. This control should limit the possibility for spread of disease from cultured to wild stocks. Many of the disease problems in culture operations are a function of holding fish at high densities (relative to natural conditions) in restricted areas. Transmission of disease under monoculture conditions can occur very rapidly. Publicly and privately supported research for improved disease diagnosis, medicines, and vaccines has resulted in a reduction of disease problems. The documentation of transfers of fish and fish eggs across national boundaries and even between hatcheries and farms in the same water drainage area serves to assist in controlling the spread of disease.

Despite frequently voiced fears, cold water fish farming for salmonids does not appear to involve fish pathogens that are harmful to human health. Obviously, any waters affected by sewage discharge or other pathways of pollution can have an effect on human health, independent of any fish culture operation that may be present. It is possible that salmon produced in farms may be less parasite prone than wild stocks and this would make them more attractive to raw fish afficionados.[304]

The concern about use of high doses of antibiotics in fish culture causing a buildup of resistance in the natural environment appears misplaced. Use of antibiotics has decreased due to better management of culture sites. Costs of heavy antibiotic use are prohibitive and loss to the environment represents waste in the eyes of the culturist. The dilution of the antibiotic within a short distance from the culture site quickly reduces its effectiveness and possible effects on nontarget animals — even filter feeders.[504]

H. PREDATION

Predation can be considered primarily a management problem for the salmonid culturist. Design, deployment, and maintenance of anti-avifauna nets above the culture facilities and nets and other deterrents in the water column represent the most effective response. Predation on salmonid culture becomes a policy issue when the predator species is one that is protected under law. While this may be problematic in many parts of the world, it is especially so in the United States where the Migratory Bird Treaty Act, the Endangered Species Act, and the Marine Mammal Protection Act combine to restrict the range of active harassment and lethal options that a fish culturist may take to protect operations. Under certain circumstances, the siting of an aquaculture facility in the vicinity of sensitive wildlife habitat may be precluded.[304]

I. RECREATION/VISUAL QUALITY/NOISE/ODOR

In areas with large populations of coastal residents a number of direct use conflicts may arise over salmonid culture. These relate to floating net-pen culture more than tank or pond culture on land. Even in areas where water temperatures tend to discourage in-water uses like swimming, recreational boating, and fishing, passive viewing and similar activities can be enjoyed by large numbers of participants on a seasonal basis. People develop intense interests in the water bodies around which they live and may become quite protective of them from perceived intrusion. Floating net-pen culture of salmon and shellfish is a non-traditional use of the marine space, and this has touched off considerable controversy over siting. Opponents find net-pen culture a nonconforming use. This poses policy problems for the agencies charged with development and management of fish culture and, of course, is discouraging to the prospective fish or shellfish farmer.

With this sort of opposition to fish culture, it is not surprising that the areas where there are relatively few people on long coastlines in temperate environments are also areas where net-pen culture has burgeoned.

V. EMERGING ISSUES

Besides the fairly well known policy issues treated above, there are a number of others that are worth monitoring. These include:

1. The effect of acid rain on salmonids and the possible role that culture can play,
2. The impact of global climate change on salmonid stocks and on fish culture operations,
3. The consequences of potentially vastly increased hatchery or ocean ranching types of fish culture that exceed the carrying capacity of the ocean commons,
4. The impact that evolving legal concepts, like the public trust doctrine, might have on salmonid culture, and
5. The role that fish culture plays in dealing with threatened or endangered salmonid species (however defined).

Acid rain affects large areas of Scandinavia and eastern North America. Salmonids have been affected by changes in the pH of their waters and declines have resulted in populations living in habitats with low buffering capacity. Mariculture is unlikely to be affected directly but may be affected indirectly through loss of genetic resources. Certainly any restoration program using cultured stocks will have a difficult time in establishing self-sustaining runs in degraded habitat.[505,506]

Global climate change may have differential effects on salmon stocks around the world. Salmonids are sensitive to relatively small changes in temperature and have evolved to certain hydrographic and climatological regimes that could be altered. Obviously, the possibility of sea level rise would not notably affect net-pen culture. However, changes in river flow rates and timing, estuary conditions, and upwelling systems can affect survival of salmon at various stages of their life cycles. Even relatively mild changes may have important consequences on the highly valued stocks of salmon and trout that would have repercussions in commercial and recreational harvests.[507]

An important question for salmon culture is, ''What is the carrying capacity of the oceans for salmon — can it be exceeded?'' Based on limited understanding of the marine ecosystem and the historical role of salmon therein, it would appear that we are far from saturation. Limits exist more in the terrestrial environment in the form of lost, degraded, or otherwise limited habitat. As demand increases for salmonid species, can culture activities continue to expand to help meet it — and what will be the impact on genetic resources? Already there are concerns about the competition between hatchery-reared and wild stocks.[508]

The public trust doctrine expresses the role of government to hold the rights to use common lands like river beds, seashores, etc. in trust for the public. Gradually, the full extent of the public trust doctrine is becoming clear through court cases and legal analyses. Aquatic lands that fall under state ownership are held in trust, and it is those lands, generally, that are needed for siting of salmonid culture facilities. Fish culture is among the beneficial uses that can be allowed on trust lands. Still, the application of the doctrine is on a case-by-case basis and it remains to be seen how the concept develops.[304]

Along with the questions posed by a potential superabundance of salmon comes the opposite situation where stocks of species are at risk of extinction. To what extent can salmonid culture assist in the recovery of stocks and in the maintenance of the genetic material in those stocks? To what extent is fish culture at fault for alterations in the fitness of natural stocks? It may not be possible to answer these questions, but they point out essential lines for research which will be required for the informed development of public policies to guide management of the stocks. At present, we lack clear understanding and appreciation of what needs to be done. As some indicate,[509] Pacific salmon are at the crossroads.[510]

REFERENCES

1. Stickney, R. R., Ed. *Culture of Nonsalmonid Freshwater Fishes,* CRC Press, Boca Raton, Florida, 1986, 1.
2. Smith, G. R., and R. F. Stearley. The classification and scientific names of rainbow and cutthroat trouts, *Fisheries,* 14(1), 4, 1989.
3. Robins, C. R., R. M. Bailey, C. E. Bond, J. R. Brooker, E. A. Lachner, R. N. Lea, and W. B. Scott. *A List of Common and Scientific Names of Fishes from the United States and Canada,* American Fisheries Society, Bethesda, Maryland, 1980, 1.
4. Mahnken, C. V. W. Personal communication, 1990.
5. Hart, J. L. *Pacific Fishes of Canada,* Fisheries Research Board of Canada, Ottawa, 1973.
6. Bailey, J. E. Alaska's Fishery Resources: the Pink Salmon, Fishery Leaflet 619, U.S. Fish and Wildlife Service, Washington, D.C., 1969, 1.
7. Combs, B. D., and R. E. Burrows. Threshold temperatures for the normal development of chinook salmon eggs, *Prog. Fish-Cult.,* 19, 3, 1957.
8. Merrell, T. R., Jr. Alaska's Fishery Resources: the Chum Salmon, Fishery Leaflet 632, U.S. Department of the Interior, Washington, D.C., 1970, 1.
9. Bakkala, R. G. Synopsis of Biological Data on the Chum Salmon, *Oncorhynchus keta* (Walbaum) 1792, FAO Species Synopsis No. 41, Circular 315, Washington, D.C., 1970, 1.
10. Sano, S. Salmon of the North Pacific Ocean. III. A review of the life history of North Pacific salmon. III. Chum salmon in the Far East, *Int. N. Pac. Fish. Comm. Bull.,* 18, 41, 1966.
11. Atkinson, C. E., J. H. Rose, and T. O. Duncan. Salmon of the North Pacific Ocean IV. Spawning populations of North Pacific salmon. 4. Pacific salmon in the Unites States, *Int. N. Pac. Fish. Comm. Bull.* 23, 43, 1967.
12. Downs, W. Pacific salmon in the Great Lakes, Wisconsin Sportsman, March, 1986.
13. Miller, D. J., and R. N. Lea. Guide to the Coastal Marine Fishes of California, *Calif. Dept. Fish Game Fish Bull.* 157, 1972, 1.
14. Laufle, J. C., G. B. Pauley, and M. F. Shepard. Species Profiles: Life Histories and Environmental Requirements of Coastal Fishes and Invertebrates (Pacific northwest) — Coho Salmon, U.S. Fish Wildl. Serv. Biol. Rep. 82(11.48), U.S. Army Corps of Engineers, TR EL-82-4, 1986, 1.
15. Hallock, R. J., and D. H. Fry. Five species of salmon, *Oncorhynchus,* in the Sacramento River, California, *Calif. Fish Game,* 53, 5, 1967.
16. Gribanov, V. I. Coho (*Oncorhynchus kisutch* Walb.) (General Biology), Izvestiia TINRO, 28, Fish. Res. Bd. Can. Transl. Ser. No. 370, 1948, 1.
17. Reiser, D. W., and T. C. Bjornn. Habitat requirements of anadromous salmonids, in: W. R. Meehan (Eds.), Influence of Forest and Range Management on Anadromous Fish Habitat in Western North America, U.S. Forest Service Gen. Tech. Rep. PNW-96, Pacific Northwest Forest and Range Experiment Station, Portland, Oregon, 1979, 1.
18. Tang, J., M. D. Bryant, and E. L. Brannon. Effect of temperature extremes on the mortality and development rates of coho salmon embryos and alevins, *Prog. Fish-Cult.,* 49, 167, 1987.
19. Shapovalov, L., and A. C. Taft. The life histories of the steelhead rainbow trout (*Salmo gairdneri gairdneri*) and silver salmon (*Oncorhynchus kisutch*) with special reference to Waddell Creek, California and recommendations regarding their management, Calif. Dept. Fish Game Fish. Bull. 98, 1, 1954.
20. Scott, W. B., and E. J. Crossman. Freshwater fishes of Canada, Fish. Res. Bd. Can. Bull. 184, 1973, 1.
21. Shapovalov, L., and W. Berrian. An experiment in hatching silver salmon (*Oncorhynchus kisutch*) eggs in gravel, *Trans. Am. Fish. Soc.,* 69, 135, 1940.
22. Tschaplinski, P. J. Aspects of the population biology of estuary-reared and stream-reared juvenile coho salmon in Carnation Creek: a summary of current research, in: G. F. Hartman (Ed.), *Proc. of the Carnation Creek Workshop: A Ten-year Review,* Malaspina College, Nanaimo, British Columbia, 289, 1982.
23. Murphy, M. J., J. F. Thedinga, K. V. Koski, and G. B. Grette. A stream ecosystem in an old-growth forest in southeast Alaska. V. seasonal changes in habitat utilization by juvenile salmonids, in: W. R. Meehan, T. R. Merrell, Jr., and T. A. Hanely (Eds.), *Proc. Symp. on Fish and Wildlife Relationships in Old-Growth Forests,* American Institute of Fishery Research Biologists, 1984, 89.
24. Hassler, T. J. Species Profiles: Life Histories and Environmental Requirements of Coastal Fishes and Invertebrates (Pacific Southwest) — Coho Salmon, U.S. Fish Wildl. Serv. Biol. Rep. 82(11.70), U.S. Army Corps of Engineers, Washington, D.C., TR EL-82-4, 1987, 1.
25. Donaldson, L. R., and E. L. Brannon. The use of warmed water to accelerate the production of coho salmon, *Fisheries,* 1(4), 1976, 12.

26. McDonald, J. The behavior of Pacific salmon fry during their downstream migration to freshwater and saltwater nursery areas, *J. Fish. Res. Bd. Can.*, 17, 655, 1960.
27. Wedemeyer, G. A., R. L. Saunders, and W. C. Clarke. Environmental factors limiting smoltification and early marine survival of anadromous salmonids, *Mar. Fish. Rev.*, 42, 1, 1980.
28. Sedgwick, S. D. *Salmon Farming Handbook,* Fishing News Books Ltd., Farnham, England, 1988, 1.
29. Hartman, W. L. Alaska's Fishery Resources: the Sockeye Salmon, Fishery Leaflet 636, U.S. Department of the Interior, Washington, D.C., 1971, 1.
30. Combs, B. D. Effect of temperature on development of salmon eggs, *Prog. Fish-Cult.*, 27, 134, 1965.
31. Beauchamp, B. A. Ecological Relationships of Hatchery Rainbow Trout in Lake Washington, Ph.D. dissertation, University of Washington, Seattle, 1987, 1.
32. Stickney, R. R. *Flagship: a History of Fisheries at the University of Washington,* Kendall/Hunt, Dubuque, Iowa, 1989, 1.
33. Raleigh, R. F., W. J. Miller, and P. C. Nelson. Habitat Suitability Index Models and Instream Flow Suitability Curves: Chinook Salmon, U.S. Fish Wildl. Serv. Biol. Rep. 82(10.122), 1986, 1.
34. Major, R. L., J. Ito, S. Ito, and H. Godfrey. Distribution and abundance of chinook salmon (*Oncorhynchus tshawytscha*) in offshore water of the North Pacific Ocean, *Int. North Pacific Fish. Comm. Bull.*, 38, 1, 1978.
35. Behnke, R. J. 1985. Chinook salmon, *Trout Magazine,* 26(2), 50, 1985.
36. Allen, M. A., and T. J. Hassler. Species Profiles: Life Histories and Environmental Requirements of Coastal Fishes and Invertebrates (Pacific southwest) — Chinook Salmon, U.S. Fish Wildl. Serv. Biol. Rep. 82(11.49), U.S. Army Corps of Engineers, TR EL-82-4, 1986, 1.
37. Aro, K. V. and M. P. Shepard. Pacific salmon in Canada, *Int. North Pacific Fish. Comm. Bull.*, 23, 225, 1967.
38. Fulton, L. A. Spawning Areas and Abundance of Chinook Salmon (*Oncorhynchus tshawytscha*) in the Columbia River Basin — past and present, U.S. Fish Wildl. Serv. Spec. Sci. Rep. Fish. 571, 1968, 1.
39. Meehan, W. R., and D. B. Siniff. A study of the downstream migrations of anadromous fishes in the Taku River, Alaska, *Trans. Am. Fish. Soc.*, 91, 399, 1962.
40. Burner, C. J. Characteristics of spawning nests of Columbia River salmon, U.S. Fish Wildl. Serv. Fish. Bull. 52, 96, 1951.
41. Briggs, J. C. The behavior and reproduction of salmonid fishes in a small coastal stream, *Calif. Dept. Fish Game Fish Bull.*, 94, 1, 1953.
42. Vronskiy, B. B. Reproductive biology of the Kamchatka River chinook salmon (*Oncorhynchus tshawytscha*), *J. Ichthyol.*, 12, 259, 1972.
43. Rounsefell, G. A. Fecundity of North American Salmonidae, U.S. Fish Wildl. Serv. Fish. Bull., 57, 451, 1957.
44. Leitritz, E., and R. C. Lewis. Trout and salmon culture (hatchery methods), *Calif. Dept. Fish Game Fish. Bull.*, 164, 1980, 1.
45. Piper, R. G., I. B. McElwain, L. E. Orme, J. P. McCraren, L. G. Fowler, and J. R. Leonard. Fish Hatchery Management, U.S. Fish and Wildlife Service, Washington, D.C., 1982, 1.
46. Chambers, J. S. Research Relating to Study of Spawning Grounds in Natural Areas, U.S. Army Corps of Engineers, North Pacific Division, Fish. Eng. Res. Prog., 1956, 88.
47. Alderdice, D. F., and F. P. J. Velson. Relation between temperature and incubation time for eggs of chinook salmon (*Oncorhynchus tshawytscha*), *J. Fish. Res. Bd. Can.*, 35, 69, 1978.
48. Seymour, A. H. Effects of Temperature upon Young Chinook Salmon, Ph.D. dissertation, University of Washington, Seattle, 1956, 1.
49. Bustard, D. R. Juvenile salmonid winter ecology in a northern British Columbia river — a new perspective, presented to American Fisheries Society North Pacific International Chapter meeting, Bellingham, WA, February 22—24, 1973, 1.
50. Gilbert, C. H. Age at maturity of the Pacific coast salmon of the genus *Oncorhynchus, U.S. Bur. Fish., Fish Bull.*, 32, 1, 1913.
51. Kjelson, M. A., P. F. Raquel, and F. W. Fisher. Life history of fall-run juvenile chinook salmon, *Oncorhynchus tshawytscha,* in the Sacramento-San Joaquin estuary, California, in: V. Kennedy (Ed.), *Estuarine Comparisons*, Academic Press, New York, 1982, 393.
52. Reimers, P. E. The Length of Residence of Juvenile Fall Chinook Salmon in Sixes River, Oregon, Oregon Fish Comm. Res. Rep., 4 1973, 1.
53. Schaffter, R. G. Fish Occurrence, Size, and Distribution in the Sacramento River Near Hood, California during 1973 and 1974, Calif. Fish Game Anad. Fish. Admin. Rep. No. 80-3, 1980, 1.
54. Slater, D. W. Winter run chinook salmon in the Sacramento River, California, with notes on winter temperature requirements during spawning, *U.S. Fish Wildl. Serv. Spec. Sci. Rep. Fish.* No. 461, 1963, 1.

55. Rutter, C. Natural history of the quinnat salmon, *U.S. Fish Comm. Bull.*, 22, 65, 1904.
56. Rich, W. H. Early life history and seaward migration of chinook salmon in the Columbia and Sacramento Rivers, *U.S. Bur. Fish. Bull.*, 37, 1, 1920.
57. Eddy, S., and J. C. Underhill. *Northern Fishes,* University of Minnesota Press, Minneapolis, 1974, 1.
58. Barnhart, R. A. Species Profiles: Life Histories and Environmental Requirements of Coastal Fishes and Invertebrates (Pacific Southwest) — Steelhead, U.S. Fish and Wildlife Service, Biol. Rep. 82(11.60), U.S. Army Corps of Engineers, TR EL-82-4, Washington, D.C., 1986, 1.
59. Withler, I. L. Variability in life history characteristics of steelhead trout (*Salmo gairdneri*) along the Pacific coast of North America, *J. Fish. Res. Bd. Can.*, 23, 365, 1966.
60. Salo, E. O. Anadromous fishes, in: *Salmonid Management, Trout, (winter Suppl.),* 1974, 12.
61. Washington, P. M. Occurrence on the high seas of a steelhead trout in its ninth year, *Calif. Fish Game,* 56, 312, 1970.
62. Maher, F. P., and P. A. Larkin. Life history of steelhead trout of the Chilliwack River, British Columbia, *Trans. Am. Fish. Soc.*, 84, 27, 1954.
63. Sumner, F. H. Age and growth of steelhead trout, *Salmo gairdneri* Richardson, caught by sport and commercial fishermen in Tillamook County, Oregon, *Trans. Am. Fish. Soc.*, 75, 77, 1945.
64. Smith, S. B. Racial Characteristics in Stocks of Anadromous Rainbow Trout, *Salmo gairdneri* Richardson, Ph.D. dissertation, University of Alberta, Edmonton, 1960, 1.
65. Smith, S. B. Reproductive isolation in summer and winter races of steelhead trout, in: T. G. Northcote (Ed.). *Symposium on Salmon and Trout in Streams*, University of British Columbia, Institute of Fisheries, Vancouver, British Columbia, 1969, 21.
66. Everest, F. H. Ecology and Management of Summer Steelhead in the Rogue River, Oregon State Game Comm. Fish. Res. Rep. No. 7, Corvallis, 1973, 1.
67. Chilcote, M. W., B. A. Crawford, and S. A. Leider. A genetic comparison of sympatric populations of summer and winter steelheads, *Trans. Am. Fish. Soc.*, 109, 203, 1980.
68. Chilcote, M. W., S. A. Leider, J. J. Loch, and R. F. Leland. Kalama River Salmonid Studies, 1982 Progress Report, Washington Department of Game Fish. Res. Rep., 83-3, Washington Department of Game, Olympia, 1983, 1.
69. Sheppard, M. F. Timing of Adult Steelhead Migrations as Influenced by Flow and Temperature in Four Representative Washington Streams, M.S. thesis, University of Washington, Seattle, 1972, 1.
70. Kesner, W. D., and R. A. Barnhart. Characteristics of fall-run steelhead trout (*Salmo gairdneri gairdneri*) of the Klamath River system with emphasis on the half-pounder, *Calif. Fish Game,* 58, 204, 1972.
71. Bulkley, R. V. Fecundity of steelhead trout, *Salmo gairdneri,* from Alsea River, Oregon, *J. Fish. Res. Bd. Can.*, 24, 917, 1967.
72. Moyle, P. B. *Inland fishes of California,* University of California Press, Berkeley. 1976, 1.
73. Smith, A. K. Development and application of spawning velocity and depth criteria for Oregon salmonids, *Trans. Am. Fish. Soc.*, 102, 312, 1973.
74. Hunter, J. W. A Discussion of Gamefish in the State of Washington as Related to Water Requirements, Wash. State Dept. of Game, Fish. Management Division Report, Washington State Department of Game, Olympia. 1973, 1.
75. Bovee, K. D. Probability of Use Criteria for the Family Salmonidae, Instream Flow Information Paper No. 4, U.S. Fish Wildl. Serv. FWS/OBS-78/07, Washington, D.C., 1978, 1.
76. Wesche, T. A., and P. A. Rechard. A Summary of Instream Flow Methods for Fisheries and Related Research Needs, Univ. Wyoming Water Resour. Res. Inst., Eisenhower Consortium Bull. 9, University of Wyoming, Casper, 1980, 1.
77. Pauley, G. B., B. M. Bortz, and M. F. Shepard. Species Profiles: Life Histories and Environmental Requirements of Coastal Fishes and Invertebrates (Pacific Northwest) — Steelhead Trout, U.S. Fish Wildl. Serv. Biol. Rep. 82(11.62), U.S. Army Corps of Engineers, TR EL-82-4, Washington, D.C., 1986, 1.24 p.
78. Bell, M. C. Fisheries Handbook of Engineering Requirements and Biological Criteria, U.S. Army Corps of Engineers Contract No. DACW 57-68-C-0086, U.S. Army Corps of Engineers, Portland, Oregon, 1973, 1.
79. Wydoski, R. S., and R. R. Whitney. *Inland Fishes of Washington,* University of Washington Press, Seattle, 1979, 1.
80. Bjornn, T. C. Embryo Survival and Emergence Studies; Job No. 5; Federal aid in fish and wildlife restoration, Job Completion Report, Project F-49-R-7, Idaho Fish and Game Dept., Boise, 1969, 1.
81. McCuddin, M. E. Survival of Salmon and Trout Embryos and Fry in Gravel-Sand Mixtures, M.S. thesis, University of Idaho, Moscow, 1977, 1.
82. Tappel, P. D., and T. C. Bjornn. A new method of relating size of spawning gravel to salmonid embryo survival, *North Am. J. Fish. Man.*, 3, 123, 1983.

83. Chapman, D. W. Food and space as regulators of salmonid populations in streams, *Am. Nat.*, 100, 345, 1966.
84. Bjornn, T. C. Survival, Production and Yield of Trout and Chinook Salmon in the Lemni River, Idaho, Bull. No. 27, College of Forestry, Wildlife and Range Sciences, University of Idaho, Moscow, 1978, 1.
85. Pollard, H. The effects of angling and hatchery trout on juvenile steelhead trout, In: J. R. Moring (Ed.), *Proc. Wild Trout — Catchable Trout Symposium,* Oregon Dept. Fish Wildlife, Corvallis, 1978, 166.
86. Folmar, L. C., and W. W. Dickhoff. The parr-smolt transformation (smoltification) and seawater adaptation in salmonids: review of selected literature, *Aquaculture,* 21, 1, 1980.
87. Conte, F. P., and H. H. Wagner. Development of osmotic and ionic regulation in juvenile steelhead trout, *Salmo gairdneri, Comp. Biochem. Physiol.*, 14, 603, 1965.
88. Fessler, J. L., and H. H. Wagner. Some morphological and biochemical changes in steelhead trout during the parr-smolt transformation, *J. Fish. Res. Bd. Can.*, 26, 2823, 1969.
89. Royal, L. A. An Examination of the Anadromous Trout Program of the Washington State Game Department, Wash. State Department of Game Final Rep., AFS-49, Olympia, 1972, 1.
90. Slatick, E., L. G. Gilbreath, and J. R. Harmon. Imprinting steelhead for homing, in: E. L. Brannon and E. O. Salo (Eds.), *Salmon and Trout Migratory Behavior Symposium,* University of Washington, Seattle, 1981, 247.
91. Wagner, H. H. 1974. Photoperiod and temperature regulation of smolting in steelhead trout (*Salmo gairdneri*), *Can. J. Zool.*, 52, 805, 1974.
92. Kerstetter, T. H., and M. Keeler. Smolting in Steelhead Trout *Salmo gairdneri* : a Comparative Study of Populations in two Hatcheries and the Trinity River, Northern California, Using Gill Na, K, ATPase Assays, Humboldt State University Sea Grant Project, HSU-SG9, Humboldt State University, Acarta, California, 1976, 1.
93. LeBrasseur, R. J. Stomach contents of salmon and steelhead trout in the northeastern Pacific Ocean, *J. Fish. Res. Bd. Can.*, 23, 85, 1966.
94. Manzer, J. I. Food of Pacific salmon and steelhead trout in the Pacific Ocean, *J. Fish. Res. Bd. Can.*, 25, 1085, 1968.
95. Pauley, G. B., K. Oshima, and K. L. Bowers. Species Profiles: Life Histories and Environmental Requirements of Coastal Fishes and Invertebrates (Pacific Northwest) — Sea-Run Cutthroat Trout, U.S. Fish Wildl. Serv. Biol. Rep. 82(11.86), U.S. Army Corps of Engineers, TR EL-82-4, Washington, D.C., 1989, 1.
96. Johnston, J. M., and S. P. Mercer. Sea-Run Cutthroat in Saltwater Pens: Broodstock Development and Extended Juvenile Rearing (with a Life History Compendium), Wash. State Game Dept. Fish. Res. Rep. AFS-57-1, 1976, 1.
97. Johnston, J. M. Life histories of anadromous cutthroat with emphasis on migratory behavior. In: E.L. Brannon and E.O. Salo (Eds.), *Salmon and Trout Migratory Behavior Symposium,* University of Washington Press, Seattle, 1982, 123.
98. Sumner, F. H. A Contribution to the Life History of the Cutthroat Trout in Oregon, Oregon State Game Commission, Salem, 1952, 1.
99. Anderson, B. G., and D. W. Narver. Fish Populations of Carnation Creek and other Barkley Sound Streams — 1974: Data Record and Progress Report, *Fish. Res. Bd. Can. Misc. Spec. Rep.* No. 1351, 1975, 1.
100. Jones, D. E. Life History Study of Sea-Run Cutthroat Trout and Steelhead Trout in Southeast Alaska, Federal aid in Fish Restoration Annual Report of Progress, 1971—1972, Alaska Dept. Fish Game Proj. F-9-4, 13(G-11-1), 1, 1972.
101. Jones, D. E. Steelhead and Sea-Run Cutthroat Trout Life History in Southeast Alaska, Anadromous Fish Studies, Annual Report of Progress, 1972—1973, Alaska Dept. Fish Game Proj. AFS-42, 14(AFS-42-1), 11, 1973.
102. Jones, D. E. Life History of Sea-Run Cutthroat Trout in Southeast Alaska, Anadromous Fish Studies Annual Report of Progress, 1973-1974, Alaska Dept. Fish Game Proj. AFS-42, 15(AFS-42-22): 15, 1974.
103. Jones, D. E. Life History of Sea-Run Cutthroat Trout in Southeast Alaska, Anadromous Fish Studies, Annual Performance Report, 1974—1975, Alaska Dept. Fish Game Proj. AFS-42, 16 (AFS-42-3-B), Section J, 23, 1975.
104. Jones, D. E. Life History of Sea-Run Cutthroat Trout in Southeast Alaska, Anadromous Fish Studies, Annual Performance Report, 1975—1976, Alaska Dept. Fish Game Proj. AFS-42, 117 (AFS-42-4-B), Section M, 29, 1976.
105. Tipping, J. M. Effect of release size on return rates of hatchery sea-run cutthroat trout, *Prog. Fish-Cult.*, 48, 195, 1986.
106. Moring, J. R., and R. L. Lantz. The Alsea Watershed Study. I. Biological Studies, Oregon Dept. Fish Wildl., Fish Res. Rep. 9, 1975, 1.

107. Jones, D. E. Development of Techniques for Enhancement of Anadromous Cutthroat Trout in Southeast Alaska, Anadromous Fish Studies, Annual Performance Report, 1976—1977, Alaska Dept. Fish Game Proj. AFS-42, 18 (AFS-42-5-BV), 1977, 28.
108. Fuss, H. J. Some life history characteristics obtained by scale analysis of sea-run cutthroat trout (*Salmo clarki clarki*) from streams near Forks, Washington, in: J. R. Moring (Ed.), *Proc. Wild Trout-Catchable Trout Symposium,* Oregon Department of Fish and Wildlife, Eugene, 1978, 127.
109. Giger, R. D. Ecology and Management of Coastal Cutthroat Trout in Oregon, Oregon State Game Comm. Fish Res. Rep. No. 6., 1972, 1.
110. Sumner, F. H. Migration of salmonids in Sand Creek, Oregon, *Trans. Am. Fish. Soc.,* 82, 139, 1953.
111. Michael, J. H. Contribution of cutthroat trout in headwater streams to the sea-run population, *Calif. Fish Game,* 69: 68, 1983.
112. June, J. A. Life History and Habitat Utilization of Cutthroat Trout (*Salmo clarki*) in a Headwater Stream in the Olympic Peninsula, Washington, M.S. thesis, University of Washington, Seattle, 1981, 1.
113. Hooper, D. R. Evaluation of the Effects of Flows on Trout Stream Ecology, California Deparment Engineering Research, Emeryville, 1973, 1.
114. Cramer, F. K. Notes on the natural spawning of cutthroat trout (*Salmo clarki clarki*) in Oregon, *Proc. Sixth Pacific Sci. Congr.* No. 3, 335, 1940.
115. Fuss, H. J. Age, Growth and Instream Movement of Olympia Peninsula Coastal Cutthroat Trout (*Salmo clarki clarki*), M.S. thesis, University of Washington, Seattle, 1982, 1.
116. Merriman, D. The effect of temperature on the development of the eggs and larvae of the cutthroat trout (*Salmo clarki clarki* Richardson), *J. Exp. Biol.,* 12, 297, 1935.
117. Snyder, G. R., and H. A. Tanner. Cutthroat Trout Reproduction in the Inlets to Trappers Lake, Colo. Fish Game Tech. Bull. 7, 1960, 1.
118. Behnke R. J., and M. Zarn. Biology and management of threatened and endangered western trout, *U.S. For. Serv. Gen. Tech. Rep. RM* 28, 1976, 1.
119. Trojnar, J. R. Ecological Evaluation of Two Sympatric Strains of Cutthroat Trout, M.S. thesis, Colorado State University, Fort Collins, 1972, 1.
120. Sekulich, P. T. Role of the Snake River cutthroat trout (*Salmo clarki* ss.) in fishery management, M.S. thesis, Colorado State University, Fort Collins, 1974, 1.
121. Dimick, R. E., and D. C. Mote. A Preliminary Survey of the Food of Oregon Trout, Oregon State College Agric. Exp. Stn. Bull. No. 323, 1934, 1.
122. Lowery, G. R. Production and food of cutthroat trout in three Oregon coastal streams, *J. Wildl. Manag.,* 30, 754, 1966.
123. Allen, K. R. Limitations on production in salmonid populations in streams, in: T.G. Northcote (Ed.), *Symposium on Salmon and Trout in Streams,* H. R. MacMillan Lectures in Fisheries, University of British Columbia, Vancouver, 1969, 3.
124. Carlander, K. D. *Handbook of Freshwater Fishery Biology,* Vol. 1, Iowa State University Press, Ames, 1969, 1.
125. Baxter, G. T., and J. R. Simon. Wyoming fishes, Wyoming Game Fish Dept. Bull. 4, 1970, 1.
126. Griffith, J. S. Utilization of invertebrate drift by brook trout (*Salvelinus fontinalis*) and cutthroat trout (*Salmo clarki*) in small streams in Idaho, *Trans. Am. Fish. Soc.,* 103, 440, 1975.
127. Glova, G. J. Management implications of the distribution and diet of sympatric populations of juvenile coho salmon and coastal cutthroat in small streams in British Columbia, Canada, *Prog. Fish-Cult.,* 46, 269, 1984.
128. McAfee, W. R. Lehonton cutthroat trout, in: A. Calhoun (Ed.), *Inland Fisheries Management,* Calif. Dept. Fish Game, Sacramento, 1966, 225.
129. Trojnar, J. R., and R. H. Behnke. Management implications of ecological segregation between two introduced populations of cutthroat trout in a small Colorado lake, *Trans. Am. Fish. Soc.,* 103, 423, 1974.
130. Hickman, T. J. Studies on Relict Populations of Snake Valley Cutthroat Trout in Western Utah 1976, U.S. Bureau of Land Management, Salt Lake City, Utah, 1977, 1.
131. Martin, D. J. Growth, food consumption, and production of cutthroat in relation to food supply and water temperature, in: J. M. Walton and D. B. Houston (Eds.), *Proc. Olympic Wild Fish Conference,* U.S. Park Service and Peninsula College, Port Angeles, Washington, 1984, 135.
132. Idyll, C. Food of rainbow, cutthroat and brown trout in the Cowichan River system, B.C., *J. Fish. Res. Bd. Can.,* 5, 448, 1942.
133. Armstrong, R. H. Age, food, and migration of sea-run cutthroat trout, *Salmo clarki,* at Eva Lake, southeastern Alaska, *Trans. Am. Fish. Soc.,* 100, 302, 1971.
134. Simenstad, C. A., and W. J. Kinney. Trophic Relationships of Outmigrating Chum Salmon in Hood Canal, Washington, 1977, Wash. State Dept. Fish. Final Rep. Contract No. 877, FRI-UW-7810, 1978, 1.

135. Horner, N., and T. C. Bjornn. Survival, Behavior, and Density of Trout and Salmon Fry in Streams, University of Idaho, For. Wildl. Exp. Stn., Contract 56, Prog. Rep. 1975, 1976, 1.
136. Sedgwick, S. D. *Salmon Farming Handbook,* Fishing News Books Ltd., Surrey, England, 1988, 1.
137. Jordan, D. S., and B. W. Evermann. *American Food and Game Fishes,* Dover Publications, Inc., New York, 1969, 1.
138. Saunders, R. L., L. B. Henderson, and B. D. Glebe. Precocious sexual maturation and smoltification in male Atlantic salmon (*Salmo salar*), *Aquaculture,* 28, 211, 1982.
139. Thorpe, J. E., C. Talbot, and C. Villarreal. Bimodality of growth and smolting in Atlantic salmon, *Salmo salar* L., *Aquaculture,* 28, 123, 1982.
140. Lundqvist, H., and G. Fridberg. Sexual maturation versus immaturity: different tactics with adaptive values in Baltic salmon (*Salmo salar* L.) male smolts, *Can. J. Zool.,* 60, 1822, 1982.
141. Parametrix, Inc. Fish Culture in Floating Net-Pens, Final Programmatic Environmental Impact Statement, Washington Department of Fisheries, Olympia, WA, 1990, 1.
142. Refstie, T., and T. Gjedrem. Hybrids between Salmonidae species, hatchability and growth rate in freshwater period, *Aquaculture,* 6, 333, 1975.
143. Loginova, G. A., and S. V. Krasnoperova. An attempt at crossbreeding Atlantic salmon and pink salmon (preliminary report), *Aquaculture,* 27, 329, 1982.
144. Raleigh, R. F., L. D. Zuckerman, and P. C. Nelson. Habitat suitability index models and instream flow suitability curves: brown trout, revised, *U.S. Fish Wildl. Serv. Biol. Rep.* 82(10.124), 1986, 1.
145. Hasler, A. D., and W. J. Wisby. Discrimination of stream odors by fish and its relation to parent stream behavior, *Am. Nat.,* 85, 223, 1951.
146. Hasler, A. D., A. T. Scholz, and R. M. Horrall. Olfactory imprinting and homing in salmon, *Am. Sci.,* 66, 347, 1978.
147. Hasler, A. D., and S. T. Kucas. Artificial imprinting of salmon in an anadromous fish hatchery, *Proc. Annu. Conf. West. Assoc. Game Fish Comm.,* 62, 448, 1982.
148. Hasler, A. D., and A. T. Scholz. *Olfactory Imprinting and Homing in Salmon,* Springer-Verlag, Berlin, 1983, 1.
149. Nordeng, J. Is the local orientation of anadromous fishes determined by pheremones?, *Nature,* 233, 411, 1971.
150. Nordeng, J. A pheromone hypothesis for homeward migration in anadromous salmonids, *Oikos,* 28, 155, 1977.
151. Brannon, E. L., and T. P. Quinn. 1989. Odor cures used by homing coho salmon, in: Brannon, E., and B. Jonsson (Eds.), *Proceedings of the Salmonid Migration and Distribution Symposium,* School of Fisheries, University of Washington, Seattle, 1989, 30.
152. Quinn, T. P., and S. Courtenay. Intraspecific chemosensory discrimination in salmonid fishes: alternative explanations, in: Brannon, E., and B. Jonsson (Eds.), *Proceedings of the Salmonid Migration and Distribution Symposium,* School of Fisheries, University of Washington, Seattle, 1989, 35.
153. Brannon, E. L., and T. P. Quinn. Field test of the pheromone hypothesis for homing by Pacific salmon, *J. Chem. Ecol.,* 16, 603, 1990.
154. Brannon, E. L. Orientation mechanisms of homing salmonids, in: E. L. Brannon and E. O. Salo (Eds.), *Salmon and Trout Migratory Behavior Symposium,* University of Washington, Seattle, 1982, 217.
155. Quinn, T. P. 1982. A model for salmon navigation on the high seas, in: E. L. Brannon and E.O. Salo (Eds.), *Salmon and Trout Migratory Behavior Symposium,* University of Washington, Seattle, 1982, 229.
156. Quinn, T. P., and C. Groot. Pacific salmon (*Oncorhynchus*) migrations: orientation versus random movement, *Can. J. Fish. Aquat. Sci.,* 41, 1319, 1984.
157. Brannon, E. L., and E. O. Salo (Eds.). 1982. *Proceedings of the Salmon and Trout Migratory Behavior Symposium,* School of Fisheries, University of Washington, Seattle, 1982, 1.
158. Brannon, E. L., and B. Jonsson (Eds.). *Proceedings of the Salmonid Migration and Distribution Symposium,* School of Fisheries, University of Washington, Seattle, 1989, 1.
159. Conte, F. P. Salt secretion, in: W. S. Hoar and D. J. Randall (Eds.). *Fish Physiology,* Vol. 1, Academic Press, New York, 1969, 241.
160. National Marine Fisheries Service. Saltwater Adaptation of Coho Salmon, Spring and Fall Chinook Salmon, and Steelhead. A Study to Assess Status of Smoltification and Fitness for Ocean Survival of Chinook, Coho, and Steelhead, U.S. Nat. Mar. Fish. Serv. Coastal Zone and Estuarine Div. Annu. Rep. for FY 1978-79, Project 817, 1979, 1.
161. Zaugg, W. S. Relationships between smolt indices and migration in controlled and natural environments, in: E. L. Brannon and E. O. Salo (Eds.), *Salmon and Trout Migratory Behavior Symposium,* University of Washington, Seattle, 1981, 173.

162. Bern, H. A. Endocrinological studies on normal and abnormal salmon smoltification, in: P. J. Gaillard and H. H. Boer (Eds.), *Comparative Endocrinology*, Elsevier/North Holland Biomedical Press, Amsterdam, 1978, 97.
163. Clarke, W. C., and J. Blackburn. A Seawater Challenge Test to Measure Smolting of Juvenile Salmon, Fish. Mar. Serv. Res. Dev. Tech. Rep. 705, Dept. of Fisheries and Environment, Canada, 1977, 1.
164. Clarke, W. C., and Y. Nagahama. The effect of premature transfer to seawater on growth and morphology of the pituitary, thyroid, pancreas and interrenal in juvenile coho salmon (*Oncorhynchus kisutch*), *Can. J. Zool.*, 55, 1620, 1977.
165. Folmar, L. C., W. W. Dickhoff, C. V. W. Mahnken, and F. W. Waknitz. Stunting and parr-reversion during smoltification of coho salmon (*Oncorhynchus kisutch*), *Aquaculture*, 28, 91, 1982.
166. Woo, N. Y. S., H. A. Bern, and R. S. Nishioka. Changes in body composition associated with smoltification and premature transfer to seawater in king salmon, *J. Fish Biol.*, 13, 421, 1978.
167. Kennedy, W. A., C. T. Shoop, W. Griffioen, and A. Solmie. The 1975 Crop of Salmon Reared on the Pacific Biological Station Experimental Fish Farm, Fish. Mar. Serv. Res. Dev. Tech. Rep. 665, Dept. of Fisheries and Environment, Canada, 1976, 1.
168. Mahnken, C, E. Prentice, W. Waknitz, G. Monan, C. Sims, and J. Williams. The application of recent smoltification research to public hatchery releases: an assessment of size/time requirements for Columbia River hatchery coho, *Aquaculture*, 28, 251, 1982.
169. Huntsman, A. G., and W. S. Hoar. Resistance of Atlantic salmon to seawater, *J. Fish. Res. Bd. Can.*, 4, 409, 1939.
170. Elson, P. F. The importance of size on the change from parr to smolt in Atlantic salmon, *Can. Fish Cult.*, 21, 1, 1957.
171. Parry, G. The development of salinity tolerance in salmon, *Salmo salar* L. and some related species, *J. Exp. Biol.*, 37, 425, 1960.
172. Johnston, C. E., and J. G. Eales. Influence of body size on silvering of Atlantic salmon (*Salmo salar*) at parr-smolt transformation, *J. Fish. Res. Bd. Can.*, 27, 933, 1970.
173. Hoar, W. S. Smolt transformation: evolution, behavior, and physiology, *J. Fish. Res. Bd. Can.*, 33, 1234, 1976.
174. Saxton, A. M., W. K. Hershberger, R. N. Iwamoto. Smoltification in the net-pen culture of coho salmon: quantitative genetic analysis, *Trans. Am. Fish. Soc.*, 113, 339, 1984.
175. Wagner, H. H. Photoperiod and temperature regulation of smolting in steelhead trout (*Salmo gairdneri*), *Can. J. Zool.*, 52, 219, 1974.
176. Mahnken, C. V. W. The size of coho salmon and time of entry into seawater. I. Effects on growth and condition index, Proc. Annu. Northwest Fish Cult. Conf., Wemme, Oregon, 24, 30, 1973.
177. Wallis, J. Recommended Time, Size and Age for Release of Hatchery Reared Salmon and Steelhead Trout, Fish Commission of Oregon, Clackamas, 1968, 1.
178. Clarke, W. C., J. E. Shelbourn, and J. R. Brett. Growth and adaptation to seawater in 'underyearling' sockeye (*Oncorhynchus nerka*) and coho (*O. kisutch*) salmon subjected to regimes of constant or changing temperature and daylength, *Can. J. Zool.*, 56, 2413, 1978.
179. Johnston, C. E., and R. L. Saunders. Parr-smolt transformation of yearling Atlantic salmon (*Salmo salar*) at several rearing temperatures, *Can. J. Fish. Aquat. Sci.*, 38, 1189, 1981.
180. Brauer, E. P. The photoperiod control of coho salmon smoltification, *Aquaculture*, 28, 105, 1982.
181. Knowles, A. H., and R. H. Rines. Environmental conditioning process of accelerating and controlling the development of salmon smolt and subsequent salt water maturation, *J. World Maricult. Soc.*, 12(2), 96, 1981.
182. Dickhoff, W. W., and C. V. Sullivan. Involvement of the thyroid gland in smoltification with special reference to metabolic and development processes, in: M. J. Dadswell, R. J. Klauda, C. M. Moffit, R. L. Saunders, R. A. Rulifson, and J. E. Cooper (Eds.), *Common Strategies of Anadromous and Catadromous Fishes*, American Fisheries Society, Bethesda, Maryland, 1987, 197.
183. Dickhoff, W. W., L. C. Folmar, and A. Gorbman. Changes in plasma thyroxine during smoltification of coho salmon, *Oncorhynchus kisutch*, *Gen. Comp. Endocrin.*, 36, 229, 1978.
184. Dickhoff, W. W., and D. S. Darling. Evolution of thyroid function and its control in lower vertebrates, *Amer. Zool.*, 23, 697, 1983.
185. Eales, J. G. Thyroid function in cyclostomes and fishes, in: E. J. Barrington (Ed.), *Hormones and Evolution*, Vol. 1, Academic Press, New York, 1977, 341.
186. McBride, J. R., D. A. Higgs, U. H. M. Fagerlund, and J. T. Buckley, Thyroid and steroid hormones: potential for control of growth and smoltification in salmonids, *Aquaculture*, 28, 201, 1982.

187. Dickhoff, W. W., C. Sullivan, and C. V. W. Mahnken. Methods of measuring and controlling the parr to smolt transformation (smoltification) of juvenile salmon, in: C. J. Sindermann (Ed.), *Proceedings of the Eleventh U.S.-Japan Meeting on Aquaculture, Salmon Enhancement,* Tokyo, Japan, October 19—20, NOAA Tech. Rep. NMFS 27, 1982, 5.
188. Dickhoff, W. W., L. C. Folmar, J. L. Mighell, and C. V. W. Mahnken. Plasma thyroid hormones during smoltification of yearling and underyearling coho salmon and yearling chinook salmon and steelhead trout, *Aquaculture,* 28, 39, 1982.
189. Folmar, L. C., and W. W. Dickhoff. Evaluation of some physiological parameters as predictive indices of smoltification, *Aquaculture,* 23, 309, 1981.
190. Dickhoff, W. W., D. S. Darling, and A. Gorbman. Thyroid function during smoltification of salmonids, in: *Phylogenetic Aspects of Thyroid Hormone Actions,* (Gunma Symposium on Endocrinology), Center for Academic Publications Japan, Tokyo, 19, 45, 1982.
191. Zaugg, W. S., and L. R. McClain. ATPase activity in gills of salmonids: seasonal variation and saltwater influence in coho salmon, *Oncorhynchus kisutch., Comp. Biochem. Physiol.,* 35, 587, 1970.
192. Zaugg, W. S., and L. R. McClain. Changes in gill ATPase activity associated with parr-smolt transformation in steelhead trout, coho and spring salmon, *J. Fish. Res. Bd. Can.,* 29, 167, 1972.
193. Ewing, R., J. Johnson, H. Pribble, and J. Lichatowich. Temperature and photoperiod effects on gill (Na-K)-ATPase activity in chinook salmon (*Oncorhynchus tshawytscha*), *J. Fish. Res. Bd. Can.,* 36, 1347, 1979.
194. Zaugg, W. S., E. F. Prentice, and F. W. Waknitz. Importance of river migration to the development of seawater tolerance in Columbia River anadromous salmonidae, *Aquaculture,* 51, 33, 1985.
195. Dickhoff, W. W., C. V. W. Mahnken, W. S. Zaugg, F. W. Waknitz, M. G. Bernard, and C. V. Sullivan. Effects of temperature and feeding on smolting and seawater survival of Atlantic salmon (*Salmo salar*), *Aquaculture,* 82, 93, 1989.
196. MacCrimmon, H. R. World Distribution of rainbow trout (*Salmo gairdneri*), *J. Fish. Res. Bd. Can.* 28, 663, 1971.
197. Miller, R. J., and E. L. Brannon. The origin and development of life history patterns in Pacific salmonids, In: E.L. Brannon and E.O. Salo (Eds.), *Salmon and Trout Migratory Behavior Symposium, 1981,* University of Washington, Seattle, 1982, 296.
198. Brannon, E. L. Mechanisms stabilizing salmonid fry emergence timing, in: H. Smith, L. Margolis, and C. Wood (Eds.), *Sockeye Salmon (Oncorhynchus nerka) Population Biology and Future Management,* Canadian Special Publications in Fisheries and Aquatic Science, 96, 120, 1987.
199. Northcote, T. G. Migratory strategies and production in freshwater species. in: S.D. Gerking (Ed.), *Ecology of Freshwater Fish Production,* Blackwell Science Publ., Oxford, 1978, 326.
200. Brannon, E. L. 1972. Mechanisms controlling migration of sockeye salmon fry, *Int. Pac. Sal. Fish. Comm. Bull.,* 21, 1, 1972.
201. Anonymous. U.S. Fish Manual, Artificial Propagation of the Salmons of the Pacific Coast, U.S. Commission of Fish and Fisheries, Government Printing Office, Washington, D.C., 1903, 1.
202. Falconer, D. S. 1967. *Introduction to Quantitative Genetics,* the Ronald Press Company, New York, 1967, 1.
203. Groot, C. 1982. Modifications on a Theme: A perspective on migratory behavior of Pacific salmon, in: E. L. Brannon and E. O. Salo (Eds.), *Salmon and Trout Migratory Behavior Symposium, 1981,* University of Washington, Seattle, 1982, 1.
204. Bams, R. A. 1976. Survival and propensity as affected by presence or absence of locally adapted parental genes in two transplanted populations of pink salmon (*Oncorhynchus gorbuscha*). *J. Fish. Res. Bd. Can.,* 33, 2716, 1976.
205. Campton, D. E., and J. M. Johnston. Electrophorectic evidence for a genetic admixture of native and nonnative rainbow trout in the Yakima River, Washington. *Trans. Am. Fish. Soc.,* 114, 782, 1985.
206. Gordon, J. W., G. A. Scangos, D. J. Plotkin, J. A. Barbosa, and F. H. Ruddle. Genetic transformation of mouse embryos by microinjection of purified DNA, *Proc. Natl. Acad. Sci. U.S.A.,* 77, 7380, 1980.
207. Hammer, R. E., V. G. Pursel, C. E. Rexroad, Jr., R. J. Wall, D. J. Bolt, K. M. Ebert, R. D. Palmiter, and R. L. Brinster. Production of transgenic rabbits, sheep and pigs by microinjection, *Nature (Lond.),* 315, 680, 1985.
208. Zhu, Z., G. Li, L. He, and S. Chen. Novel gene transfer into the fertilized eggs of gold fish (*Carassius auratus* L. 1758), *Z. Angew. Ichthyol.,* 1, 31, 1985.
209. Brem, G., B. Brenig, G. Horstgen-Schwark, and E.-L. Winnacker. Gene transfer in tilapia (*Oreochromis niloticus*), *Aquaculture,* 68, 209, 1988.
210. Dickison, J. A. Some Effects of 17 alpha-Methyltestosterone on Rainbow Trout. M.S. Thesis, University of Washington. Seattle, 1985, 1.

211. Thorgaard, G. H., M. E. Jazwin, and A. R. Stier. Polyploidy induced by heat shock in rainbow trout. *Trans. Am. Fish. Soc.*, 110, 546, 1981.
212. Hurley, D. A., and K. C. Fisher. The structure and development of the external membrane in young eggs of the book trout, *Salvelinus fontinalis, Can. J. Zool.* 44, 173, 1966.
213. Brannon, E. L. Influence of physical factors on the development and weight of sockeye salmon embryos and alevins, *Int. Pac. Sal. Fish. Comm. Progr. Rep.* 12, 1965, 1.
214. Klontz, G. W. Concept and methods of intensive aquaculture. Department of Fish and Wildlife Research, University of Idaho, Moscow, 1990, 1.
215. Peterson, H. H. Smolt rearing methods, equipment and techniques used successfully in Sweden, in: M.W. Carter (Ed.), *International Atlantic Salmon Workshop, 1971,* Special Publ. Series, Vol. 2, The International Atlantic Salmon Foundation, New York, 1971, 32.
216. Besner, M. Endurance Training: An Affordable Rearing Strategy to Increase Food Conversion Efficiency, Stamina, Growth and Survival of Coho Salmon Smolts (*Oncorhynchus kisutch*), Ph.D. thesis. University of Washington, Seattle, 1980, 1.
217. Thompson, R. B. Effects of Predator Avoidance Conditioning on the Post-Release Survival Rate of Artificially Propagated Salmon, Ph.D. dissertation, University of Washington, Seattle, 1966, 1.
218. Wedemeyer, G., and J. Wood. Stress as a Predisposing Factor in Fish Disease, U.S. Fish and Wildlife Service, Fish Disease Leaflet No. 38, 1974, 1.
219. Willoughby, H. A method for calculating carrying capacities of hatchery troughs and ponds, *Prog. Fish-Cult.*, 30, 173, 1968.
220. Wedemeyer, G. Some physiological consequences of handling stress in juvenile coho salmon (*Oncorhynchus kisutch*) and steelhead trout (*Salmo gairdneri*), *J. Fish. Res. Bd. Can.*, 29, 1780, 1972.
221. Reisenbichler, R. R., and J. D. McIntyre. Genetic differences in growth and survival of juvenile hatchery and wild steelhead trout, *Salmo gairdneri, J. Fish. Res. Bd. Can.*, 34, 123, 1977.
222. Piggins, D. J., and C. P. R. Mills. Comparative aspects of the biology of naturally produced and hatchery-reared Atlantic salmon smolts (*Salmo salar* L.), *Aquaculture,* 45, 321, 1985.
223. Altukhov, Yu. P., and E. A. Salmenkova, Applications of the stock concept to fish populations in the USSR, *Can. J. Fish. Aquat. Sci.*, 38, 1591, 1981.
224. Harlan, J. R. Who's in charge here?, *Can. J. Fish. Aquat. Sci.*, 38, 1459, 1981.
225. Ricker, W. E. Hereditary and environmental factors affecting certain salmonid populations, in: P. A. Larkin and R. C. Simon (Eds.), *The Stock Concept of Pacific Salmon,* H. R. MacMillan Lectures in Fisheries, University of British Columbia, Vancouver, Canada, 1972, 19.
226. Templeton, A. R. Coadaptation and outbreeding depression. in: M. E. Soulé (Ed.), *Conservation Biology: Science of Scarcity and Diversity,* Sinaur Associates, Sunderland, Netherlands, 1986, 105.
227. Hindar, K., N. Ryman, and G. Stahl. Genetic differentiation among local populations and morphotypes of Arctic char, *Salvelinus alpinus, Biol. J. Linn. Soc. Lond.*, 27, 269, 1986.
228. Parkinson, E. A. Genetic variation in populations of steelhead trout (*Salmo gairdneri*) in British Columbia, *Can. J. Fish. Aquat. Sci.*, 41, 1412, 1984.
229. Mayr, E. *Animal Species and Evolution,* Harvard University Press, Cambridge, Massachusetts, 1981, 1.
230. Calaprice, J. R. Production and genetic factors in managed salmonid populations, In: T. G. Northcote (Ed.), *Symposium on Salmon and Trout in Streams,* H. R. McMillan Lectures in Fisheries, University of British Columbia, Vancouver, Canada, 1969, 377.
231. Scudder, G. G. E. The adaptive significance of marginal populations: a general perspective, in: C. D. Levings, L. B. Holtby, and M. A. Henderson (Eds.), *Proceedings of the National Workshop on Effects of Habitat Alteration on Salmonid Stocks,* Can. Spec. Publ. Fish. Aquat. Sci., 105, 1989, 180.
232. Hynes, J. D., E. G. Brown, Jr., J. G. Helle, N. Ryman, and D. A. Webster. Guidelines for the culture of fish stocks for resource management, *Can. J. Fish. Aquat. Sci.*, 38, 1867, 1981.
233. Taylor, E. B., and J. D. McPhail. Variation in body morphology among British Columbia populations of coho salmon, *Oncorhynchus kisutch, Can. J. Fish. Aquat. Sci.*, 43, 2020, 1985.
234. Rosenau, M. L., and J. D. McPhail. Inherited differences in agonistic behavior between two populations of coho salmon, *Trans. Am. Fish. Soc.*, 116, 646, 1987.
235. MacLean, J. A., and D. O. Evans. The stock concept, discreteness of fish stocks, and fisheries management, *Can. J. Fish. Aquat. Sci.*, 38, 1889, 1981.
236. Shields, W. M. *Philopatry, Inbreeding, and the Evolution of Sex,* State University of New York Press, Albany, 1982, 1.
237. Endler, J. A. *Geographic Variation, Speciation, and Clines,* Princeton University Press, New Jersey, 1977, 1.
238. Carson, H. L. The genetics of speciation at the diploid level, *Am. Nat.*, 109, 83, 1975.
239. Wade, M. G. The Relative Effects of *Ceratomyxa shasta* on Crosses of Resistant and Susceptible Stocks of Summer Steelhead, M.S. thesis, Oregon State University, Corvallis, 1987, 1.

240. Chilcote, M. W., S. A. Leider, and J. J. Loch. Differential reproductive success of hatchery and wild summer-run steelhead under natural conditions, *Trans. Am. Fish. Soc.*, 115, 726, 1986.
241. Nickelson, T. E., M. F. Solazzi, and S. L. Johnson. Use of hatchery coho salmon (*Oncorhynchus kisutch*) presmolts to rebuild wild populations in Oregon coastal streams, *Can. J. Fish. Aquat. Sci.*, 43, 2443, 1986.
242. Wales, J. H. Relative survival of hatchery and wild trout, *Prog. Fish-Cult.*, 16, 125, 1954.
243. Chapman, D. W. Aggressive behaviour in juvenile coho salmon as a cause of emigration, *J. Fish. Res. Bd. Can.*, 19, 1047, 1962.
244. Enright, J. T. Climate and population regulation: the biogeographer's dilemma, *Oecologia*, 24, 295, 1976.
245. Reisenbichler, R. R. Outplanting: potential for harmful genetic change in naturally spawning salmonids, in: J. M. Walton and D. B. Houston (Eds.), *Proceedings of the Olympic Wild Fish Conference*, Peninsula College, Port Angeles, Washington, 1984, 33.
246. McIntyre, J. D., and R. R. Reisenbichler. A model for selecting harvest fraction for aggregate populations of hatchery and wild anadromous salmonids, in: R. H. Stroud (Ed.), *Fish Culture and Fisheries Management*, American Fisheries Society, Bethesda, Maryland, 1986, 170.
247. Bardach, J. E., J. G. Ryther, and W. O. McLarney. *Aquaculture, the Farming and Husbandry of Freshwater and Marine Organisms*, John Wiley & Sons, New York, 1972, 1.
248. Senn, H., J. Mack, and L. Rothfus. Compendium of Low-Cost Pacific Salmon and Steelhead Trout Production Facilities and Practices in the Pacific Northwest, available from Bonneville Power Administration, Division of Fish and Wildlife, Public Information Office, P.O. Box 3621, Portland, Oregon 97208, 1984, 1.
249. Schuck, H. A. Survival of hatchery trout in streams and possible methods of improving the quality of hatchery trout, *Prog. Fish-Cult.*, 10, 3, 1948.
250. Reisenbichler, R. R. Relation between distance transferred from natal stream and recovery rate for hatchery coho salmon, *N. Am. J. Fish. Man.*, 8, 172, 1988.
251. Allendorf, F. W., and N. Ryman. Genetic management of hatchery stocks, in: N. Ryman and F. Utter (Eds.), *Population Genetics and Fishery Management*, University of Washington Press, Seattle, 1987, 141.
252. Frankel, O. H., and M. E. Soulé. *Conservation and Evolution*, Cambridge University Press, Cambridge, 1981, 1.
253. Tave, D. *Genetics for Fish Hatchery Managers*, AVI Publishing Westport, CT, 1986, 1.
254. Wright, S. *The Theory of Gene Frequencies, Evolution and the Genetics of Populations*, Vol. 2, Chicago University Press, Chicago, 1969, 1.
255. Franklin, I. R. Evolutionary change in small populations, in: M. E. Soulé and B. A. Wilcox (Eds.), *Conservation Biology: An Evolutionary-Ecological Perspective*, Sinauer Associates, Sunderland, Massachusetts, 1980, 135.
256. Elliott, D. G., R. J. Pascho, and G. L. Bullock. Developments in the control of bacterial kidney disease of salmonid fishes, *Dis. Aquat. Org.*, 6, 201, 1989.
257. Gharrett, A. J., and S. M. Shirley. A genetic examination of spawning methodology in a salmon hatchery, *Aquaculture*, 47, 245, 1985.
258. Withler, R. E. Genetic consequences of fertilizing chinook salmon (*Oncorhynchus tshawytscha*) eggs with pooled milt, *Aquaculture*, 68, 15, 1988.
259. Wagner, H. S., R. L. Wallace, and H. J. Campbell. The seaward migration and return of hatchery-reared steelhead trout, *Salmo gairdneri* Richardson, in the Alsea River, Oregon, *Trans. Am. Fish. Soc.*, 92, 202, 1963.
260. Reisenbichler, R. R., J. D. McIntyre, and R. J. Hallock. Relation between chinook salmon, *Oncorhynchus tshawytscha*, released at hatcheries and returns to hatcheries and ocean fisheries, *Cal. Fish Game*, 68, 57, 1982.
261. Holtby, L. B., and J. C. Scrivener. Observed and simulated effects of climatic variability, clear-cut logging and fishing on the numbers of chum salmon (*Oncorhynchus keta*) and coho salmon (*O. kisutch*) returning to Carnation Creek, British Columbia, in: C. D. Levings, L. B. Holtby, and M. A. Henderson (Eds.), *Proceedings of the National Workshop on Effects of Habitat Alteration on Salmonid Stocks*, Can. Spec. Publ. Fish. Aquat. Sci., 1989, 105.
262. Banks, J. Unpublished manuscript.
263. Banks, J. Personal communication.
264. Banner, C. R., J. J. Long, J. L. Fryer, and J. S. Rohovec. Occurrence of salmonid fish infected with *Renibacterium salmoninarum* in the Pacific Ocean, *J. Fish Dis.*, 9, 273, 1986.
265. Kanayama, Y. Studies of the conditioned reflex in lower vertebrates. X. Defensive conditioned reflex of chum salmon fry in a group, *Mar. Biol.*, 2, 77, 1968.
266. Roadhouse, S., M. J. Saari, D. Roadhouse, and B. A. Pappas. Behavioral and biochemical correlates of hatchery rearing methods in lake trout, *Prog. Fish-Cult.*, 48, 38, 1986.

267. Suboski, M. S., and J. J. Templeton. Life skills training for hatchery fish: social learning and survival, *Fish. Res.*, 7, 343, 1989.
268. Leon, K. A. Effect of exercise on feed consumption, growth, food conversion, and stamina of brook trout, *Prog. Fish-Cult.*, 48, 43, 1986.
269. Shustov, Y. A., and I. L. Shchurov. Quantitative estimation of stamina of wild and hatchery-reared Atlantic salmon (*Salmo salar* L.), *Aquaculture*, 71, 81, 1988.
270. Shustov, Y. A., and I. L. Shchurov. Experimental study of the effect of young salmon, *Salmo salar* L., stamina on their feeding rates in a river, *J. Fish Biol.*, 34, 959, 1989.
271. Emlen, J. M. Personal communication.
272. Reisenbichler, R. R., and S. R. Phelps. Genetic variation in steelhead (*Salmo gairdneri*) from the north coast of Washington, *Can. J. Fish. Aquat. Sci.*, 46, 66, 1989.
273. Wohlfarth, G. W. Decline in natural fisheries — a genetic analysis and suggestion for recovery, *Can. J. Fish. Aquat. Sci.*, 43, 1298, 1986.
274. Nelson, K., and M. E. Soulé. Genetical conservation of exploited fishes, in: N. Ryman and F. Utter (Eds.), *Population Genetics and Fishery Management*, University of Washington Press, Seattle, 1987, 345.
275. Emlen, J. M., R. R. Reisenbichler, A. M. McGie, and T. E. Nickelson. Density dependence at sea for coho salmon *Oncorhynchus kisutch*, *Can. J. Fish. Aquat. Sci.*, 1990, in press.
276. Reisenbichler, R. R., and J. D. McIntyre. Requirements for integrating natural and artificial production of anadromous salmonids in the Pacific Northwest, in: R. H. Stroud (Ed.), *Fish Culture in Fisheries Management*, American Fisheries Society, Bethesda, Maryland, 1986, 365.
277. Novotny, A. J. Net-pen culture of Pacific salmon in marine waters, in: T. Nosho, R. Nakatani, E. Brannon, and E. Salo (Eds.), Workshop on salmonid aquaculture, Washington Sea Grant Report WSG-WO-74-1, Seattle, 7, 1974.
278. Hunter, C. J., and W. E. Farr. Large floating structure for holding adult Pacific salmon (*Oncorhynchus* spp.), *J. Fish. Res. Bd. Can.*, 27, 947, 1970.
279. Novotny, A. J. Net-pen culture of Pacific salmon in marine waters, *Mar. Fish. Rev.*, 37, 36, 1975.
280. Boydstun, L. B., and J. S. Hopelain. Cage rearing of steelhead rainbow trout in a freshwater impoundment, *Prog. Fish-Cult.*, 39, 70, 1977.
281. Tucker, C. S. (Ed.). *Channel Catfish Culture*, Elsevier, New York, 1985, 1.
282 Mahnken, C. V. W., A. J. Novotny, and T. Joyner. Salmon mariculture potential assessed, *Am. Fish Farmer & World Aquaculture News*, 2(1), 12, 1970.
283. Novotny, A. J., and C. V. W. Mahnken. Farming Pacific salmon in the sea, Part. 1, *Fish Farming Ind.*, 2(5), 6, 1971.
284. Novotny, A. J., and C. V. W. Mahnken. Farming Pacific salmon in the sea, Part. 2, *Fish Farming Ind.*, 3(1), 19, 1972.
285. Lindsay, C. E. Salmon farming in Washington moves closer to industry status, *Aquaculture Mag.*, 6(3), 20, 1980.
286. Aiken, D. E. The B.C. salmon culture industry, *Bull. Aquaculture Assoc. Canada*, 86-3, 4, 1986.
287. Boeuf, G., and Y. Harache. Salmonid marine rearing in France, *J. World Maricult. Soc.*, 14, 246, 1984.
288. Egan, D., and A. Kenney. Salmon farming in British Columbia, *World Aquaculture*, 21(2), 6, 1990.
289. Chettleburgh, P. West coast shake-out, Supply and demand tremors rattle B.C. growers, *Canadian Aquaculture*, September/October: 21, 1989.
290. Needham, T. Farmed salmon, where will we be in the year 2000, *Canadian Aquaculture*, May/Jun 1990, 54.
291. Boyce, J. A. Use conflicts and floating aquaculture in Puget Sound, Washington Department of Ecology, Olympia, 1988, 1.
292. Weston, D. P. 1986. *The Environmental Effects of Floating Mariculture in Puget Sound*, University of Washington, Seattle, 1986, 1.
293. Kennedy, W. A. A Handbook on Rearing Pan-Size Pacific Salmon Using Floating Seapens, Fisheries and Marine Service Industry Report No. 107, Pacific Biological Station, Nanaimo, British Columbia, 1978, 1.
294. Sutterlin, A. M., and S. P. Merrill. Norwegian Salmonid Farming, Tech. Rep. 779, Department of Fisheries and Environment, St. Andrews, New Brunswick, Fisheries and Marine Service, 1978, 1.
295. Leavens, K. An Overview of Netpen Rearing as an Enhancement Technique, prepared for the Department of Fisheries and Oceans, Vancouver, B.C., 1983, 1.
296. Nyegaard, L. 1973. Coho Salmon Farming in Puget Sound, Washington State University, Cooperatative Extension Service, College of Agriculture Bull., Pullman, Washington, 647, 1973, 1.
297. STOWW. Salmon Pen Culture at Squaxin Island Indian Reservation, prepared by the Small Tribes Organization of Western Washington, Federal Way, Washington for the Economic Development Administration, U.S. Office of Economic Opportunity and the Bureau of Indian Affairs. Washington, D.C., 1974, 1.

298. Science Applications International Corporation. *Recommended Interim Guidelines for the Management of Salmon Net-Pen culture in Puget Sound,* Science Applications International Corporation, Bellevue, Washington, 1986, 1.
299. Saxton, A. M., R. N. Iwamoto, and W. K. Hershberger. Smoltification in the net-pen culture of accelerated coho salmon, *Oncorhynchus kisutch* Walbaum: prediction of saltwater performance, *J. Fish Biol.* 22, 363, 1983.
300. Teskeredzic, E., Z. Teskeredzic, M. Tomec, D. Margus, and M. Hacmanjek. Culture of coho salmon (*Oncorhynchus kisutch*) and rainbow trout (*Salmo gairdneri*) in the Adriatic Sea, *Bull. Aquaculture Assoc. of Can.,* 88-4, 19, 1988.
301. Brett, J. R. Tank experiments on the culture of pan-sized sockeye (*Oncorhynchus nerka*) and pink salmon (*O. gorbuscha*) using environmental control, *Aquaculture,* 4, 341, 1974.
302. Wertheimer, A. C., W. R. Heard, and R. M. Martin. Culture of sockeye salmon (*Oncorhynchus nerka*) smolts in estuarine net pens and returns of adults from two smolt releases, *Aquaculture,* 32, 373, 1983.
303. Chaney, D. Personal communication, 1989.
304. Washington Department of Fisheries, *Fish Culture in Floating Net Pens,* Final Programmatic Environmental Impact Statement, Olympia, Washington, January 1990, 1.
305. Halver, J. E. Nutrition of salmonid fishes. III. Water-soluble vitamin requirements of chinook salmon, *J. Nutrition,* 62, 225, 1957.
306. Hardy, R. W., Diet preparation, In: J.E. Halver (Ed.)., *Fish Nutrition,* 2nd ed., 475, 1989.
307. Moring, J. R., and K. A. Moring. Succession of net biofouling material and its role in the diet of pen-cultured chinook salmon, *Prog. Fish-Cult.,* 37, 27, 1975.
308. Hardy, R. W. Personal communication, 1990.
309. Halver, J. E., D. C. DeLong, and E. T. Mertz. Threonine and lysine requirements of chinook salmon, *Fed. Proc. Fed. Am. Soc. Exp. Biol.,* 17, 1873, 1958.
310. Akiyama, T., S. Arai, T. Murai, and T. Nose. Threonine, histidine, and lysine requirements of chum salmon fry, *Bull. Jpn. Soc. Sci. Fish.,* 51, 635, 1985.
311. Halver, J. E., D. C. DeLong, and E. T. Mertz. Methionine and cystine requirements of chinook salmon, *Fed. Proc. Fed. Am. Soc. Exp. Biol.,* 18, 2076, 1959.
312. Fowler, L. G. Substitution of soybean and cottonseed products for fish meal in diets fed to chinook and coho salmon, *Prog. Fish-Cult.,* 42, 86, 1980.
313. Yu, T. C., and R. O. Sinnhuber. Effects of dietary n-3 and n-6 fatty acids on growth and feed onversion efficiency of coho salmon (*Oncorhynchus kisutch), Aquaculture,* 16, 31, 1979.
314. Takeuchi, T., T. Watanabe, and T. Nose. Requirement for essential fatty acids of chum salmon (*Oncorhynchus keta*) in freshwater environment, *Bull. Jpn. Soc. Sci. Fish.,* 45, 1319, 1979.
315. Donanjh, B. S., D. A. Higgs, M. D. Plotnikoff, J. R. Markert, and J. T. Buckley. Preliminary evaluation of canola oil, pork lard, and marine lipid singly and in combination as supplemental dietary lipid sources for juvenile fall chinook salmon (*Oncorhynchus tshawytscha*), *Aquaculture,* 68, 325, 1988.
316. National Research Council. *Nutrient Requirements of Coldwater Fishes,* National Academy of Sciences Press, Washington, D.C., 1981, 1.
317. Watanabe, T., A. Murakami, L. Takeuchi, T. Nose, and C. Ogino. Requirement of chum salmon held in freshwater for dietary phosphorus, *Bull. Jpn. Soc. Sci. Fish.,* 46, 361, 1980.
318. Lall, S. P. The minerals, in: J. E. Halver (Ed.), *Fish Nutrition,* 2nd ed., Academic Press, New York, 1989, 219.
319. Shearer, K. D. Dietary potassium requirement of juvenile chinook salmon, *Aquaculture,* 73, 119, 1988.
320. Fowler, L. G., and R. E. Burrows. The Abernathy salmon diet, *Prog. Fish-Cult.,* 33, 67, 1971.
321. Spinelli, J., and C. Mahnken. Carotenoid deposition in pen-reared salmonids fed diets containing oil extracts of red crab (*Pleuroncodes planipes*), *Aquaculture,* 13, 213, 1978.
322. Torrissen, O. J., R. W. Hardy, and K. D. Shearer. Pigmentation of salmonids — carotenoid deposition and metabolism, *Rev. Aquat. Sci.,* 1, 209, 1989.
323. Johnstone, R., T. H. Simpson, and A. F. Youngson. Sex reversal in salmonid culture. *Aquaculture,* 13, 115, 1978.
324. McBride, J. R., and U. H. M. Fagerlund. The use of 17α-methyltestosterone for promoting weight increases in juvenile Pacific salmon, *J. Fish. Res. Bd. Can.,* 30, 1099, 1973.
325. McBride, J. R., and U. H. M. Fagerlund. Sex steroids as growth promoters in the cultivation of juvenile coho salmon (*Oncorhynchus kisutch*), *Proc. World Maricult. Soc.,* 7, 145, 1976.
326. Higgs, D. A., U. H. M. Fagerlund, J. R. McBride, H. M. Dye, and E. M. Donaldson. Influence of bovine growth hormone, 17α-methyltestosterone, and L-thyroxine on growth of yearling coho salmon (*Oncorhynchus kisutch*), *Can. J. Zool.,* 55, 1048, 1977.
327. Matty, A. J., and I. R. Cheema. The effect of some steroid hormones on growth and protein metabolism of rainbow trout, *Aquaculture,* 14, 163, 1978.

328. Yu, T. C., R. O. Sinnhuber, and J. D. Hendricks. Effect of steroid hormones on the growth of juvenile coho salmon (*Oncorhynchus kisutch*), *Aquaculture,* 16, 351, 1979.
329. Fagerlund, U. H. M., and J. R. McBride. Growth increments and some flesh and gonad characteristics of juvenile coho salmon rccciving dicts supplemented with 17α-methyltestosterone, *J. Fish. Biol.* 7, 305, 1975.
330. Allee, B. J. The status of saltwater maturation of coho salmon (*Oncorhynchus kisutch*) at Oregon Aqua-Foods, Inc., in: T. Nosho (Ed.), *Salmonid Broodstock Maturation,* Washington Sea Grant, Seattle, Washington, 1981, 1.
331. Allee, B. J., and B. K. Suzumoto. Maturation studies of the 1980 brood year coho salmon (*Oncorhynchus kisutch*) at Oregon Aqua-Foods, Inc., in: T. Nosho (Ed.), *Salmonid Broodstock Maturation,* Washington Sea Grant, Seattle, Washington, 1981, 71.
332. Sower, S. A., and C. B. Schreck. Steroid and thyroid hormones during sexual maturation of coho salmon (*Oncorhynchus kisutch*) in seawater or fresh water, *Gen. Comp. Endocrinol.,* 47, 42, 1982.
333. Wertheimer, A. Maturation success of coho salmon and pink salmon held under different salinity regimes, in: T. Nosho (Ed.), *Salmonid Broodstock Maturation,* Washington Sea Grant, University of Washington, Seattle, 1981, 75.
334. Wertheimer, A. Use of estuarine netpens for holding returning broodstock. in: T. Nosho (Ed.), *Salmonid Broodstock Maturation,* Washington Sea Grant, University of Washington, Seattle, 1981, 9.
335. Heggelund, P. O. Washington issues and potential of salmon farming, *Aquaculture Today,* Autumn, 1989, 10.
336. Moring, J. R. Documentation of unaccounted-for-losses of chinook salmon from saltwater cages, *Prog. Fish-Cult.,* 51, 173, 1989.
337. Egidius, E. Diseases of salmonids in aquaculture. *Helgoländer Meeresunters.,* 37, 547, 1984.
338. Roberts, R. J., and C. J. Shepherd. *Handbook of Trout and Salmon Diseases,* 2nd ed., Fishing News Books, Farnham, U.K., 1986, 1.
339. Johnson, K. A., and D. F. Amend. Efficacy of *Vibrio anguillarum* and *Yersinia ruckeri* bacterins applied by oral and anal inbutation of salmonids, *J. Fish Dis.,* 6, 473, 1983.
340. Meyer, F. P., and R. A. Schnick. A review of chemicals used for the control of fish diseases, *Rev. Aquat. Sci.,* 1, 693, 1989.
341. Rensel, J. E., R. A. Horner, and J. R. Postel. Effects of phytoplankton blooms on salmon aquaculture in Puget Sound, Washington: initial research. *Northwest Envir. J.,* 5, 53, 1989.
342. Egan, B. D. 1990. All Dredged Up and No place to Go, *Aquaculture Assoc. Can. Bull.,* 90—1, 7, 1990.
343. Bell, G. R., W. Griffoen, and O. Kennedy. 1974. Mortalities of pen-reared salmon associated with blooms of marine algae, in: Proceedings of the Northwest Fish Culture Conference, 25th Anniversary, Seattle, Washington, 1974, 58.
344. Brett, J. R., W. Griffoen, and A. Solmie. The 1977 Crop of Salmon Reared on the Pacific Biological Station Experimental Fishfarm, Tech. Rep. 845, Department of Fisheries and the Environment, Fisheries and Marine Service, Nanaimo, Canada, 1978, 1.
345. Gaines, G., and F. J. R. Taylor. A Mariculturist's Guide to Potentially Harmful Marine Phytoplankton of the Pacific coast of North America, Information Report 10, Marine Resources Section, Fisheries Branch, British Columbia Ministry of Environment, Vancouver, 1986, 1.
346. Farrington, C. W. Mortality and Pathology of Juvenile Chinook Salmon (*Oncorhynchus tshawytscha*) Exposed to Cultures of the Marine Diatom *Chaetoceros convolutus,* M.S. thesis, University of Alaska-Southeast, Juneau, 1988, 1.
347. Rensel, J. E., and E. F. Prentice. Factors controlling growth and survival of cultured spot prawn, *Pandalus platyceros,* in Puget Sound, Washington, *Fish. Bull.,* 78, 781, 1980.
348. Forster, J. Personal communication, 1990.
349. Rensel, J. E. Phytoplankton and nutrient studies near salmon net-pens at Squaxin Island, Washington, in: Technical Appendix to the State of Washington Draft Programmatic Environmental Impact Statement: Fish Culture in Floating Net-Pens, prepared by Parametrix, Inc. for the Washington Department of Fisheries, Olympia, 1989, 1.
350. Isaksson, A. Salmon ranching: a world review, *Aquaculture,* 75, 1, 1988.
351. Jacobi, S. L. On the breeding of trout by impregnation of the ova. *Hanover Mag.,* 62, 1763.
352. Smith, G. R., and R. F. Stearlie. The classification and scientific names of rainbow and cutthroat trout, *Fisheries,* 14, 4, 1989.
353. Edwards, D. J. *Salmon and Trout Farming in Norway,* Fishing News Books, Ltd., Surrey, England, 1978, 1.
354. Larson, P. O. Smolt rearing and the Baltic salmon fishery, in: J. E. Thorpe (Ed.), *Salmon Ranching,* Academic Press, London, 1980, 157.

355. Eskelinen, U., and L. O. Eriksson. Havbeiting-Muligheter i Avlsarbeide, Preliminär Östersjöutredning, Havbeiteudvalget (unpubl.), 1987, 1.
356. Piggins, D. J. Salmon ranching in Ireland, in: J. E. Thorpe (Ed.), *Salmon Ranching*, Academic Press, London, 1980, 187.
357. Saunders, R. L. Sea ranching of salmon in Canada, in: C. Eriksson, M. P. Ferranti, and P. O. Larsson (Eds.), Sea Ranching of Atlantic Salmon, Proceedings of a COST 46/4 Workshop, 26—29 October 1982, Lisbon, EEC Publication, 1982, 5.
358. Reinert, A. Atlantic salmon and plans for salmon ranching in the Faroe Islands, in: C. Eriksson, M. P. Ferranti, and P. O. Larsson (Eds.), *Sea Ranching of Atlantic Salmon*, Proceedings of a COST 46/4 Workshop, 26—29 October 1982, Lisbon, EEC Publication, 1982, 81.
359. Hansen, L. P. Salmon ranching in Norway, in: C. Eriksson, M. P. Ferranti, and P. O. Larsson (Eds.), *Sea Ranching of Atlantic Salmon*, Proceedings of a COST 46/4 Workshop, 26—29 October 1982, Lisbon, EEC Publication, 1982, 95.
360. Thorpe, J. E., Salmon ranching in Britain, In: C. Eriksson, M. P. Ferranti, and P. O. Larsson (Eds.), *Sea Ranching of Atlantic Salmon*, Proceedings of a COST 46/4 Workshop, 26—29 October 1982, Lisbon, EEC Publication, 1982, 43.
361. Hanssen, O. Norwegian fish farming, Special edition of *Norsk Fiskeoppdrett*, 8 (August), 1987, 1.
362. Gjerset, F. Salmon farming in Norway, present status and outlook into the 1990s, in: W. J. McNeil (Ed.), *Salmon Production, Management, and Allocation, Biological, Economic and Policy Issues*, Oregon State University Press, Corvallis, 1988, 33.
363. Hansen, T., and D. Moller. Klekking av laks i kunstgress, *Norsk Fiskeoppdrett*, 12 (December), 4, 1983.
364. Beveridge, M. C. M. *Cage aquaculture*, Fishing News Books Ltd., Surrey, England, 353, 1987.
365. Gunnes, K. Sjöanlegg, in: T. Gjedrem (Ed.), *Oppdrett Av Laks og Aure*, Landbruksforlaget, Oslo, 132, 1979.
366. Hanssen O. Lakseoppdrett i Skottland, *Norsk Fiskeoppdrett*, 5/6, May/Jun, 1981, 1.
367. Joyce, J. Meeting the Atlantic challenge, how Ireland leads the world in offshore technology, *Aquaculture Ireland*, No. 37, May/June, 1, 1988.
368. Hanssen, O. MOWI over til merer, *Norsk Fiskeoppdrett*, 10 (October), 1, 1987.
369. Braaten, B., and I. Högoy. Produksjon av matfisk, in: O. Ingebrigtsen (Ed.), *Akvakultur, Oppdrett av laksefisk*, NKS-Forlaget, Oslo, 146, 1982.
370. Needham, T. Sea water cage culture of salmonids, in: L. Laird and T. Needham (Eds.), *Salmon and Trout Farming*, Ellis Horwood Series in Aquaculture and Fisheries Support, Chichester, 1988, 117.
371. Halver, J. E. The Vitamins, in: J. E. Halver (Ed.), *Fish Nutrition*, Academic Press, New York, 1989, 31.
372. Wilson, R. P. Amino Acids and Proteins, in: J. E. Halver (Ed.), *Fish Nutrition*, Academic Press, New York, 111, 1989.
373. Phillips, A. M., Jr., H. A. Podoliak, H. A. Poston, D. L. Livingston, H. E. Brooks, E. E. Pyle, and G. L. Hammer., The nutrition of trout, *Fish. Res. Bull. N.Y. State*, 27, 47, 1964.
374. Lovell, R. T. Diet and Fish Husbandry, in: J. E. Halver (Ed.), *Fish Nutrition*, Academic Press, New York, 1989, 549.
375. Gulbrandsen, K. E., and F. Utne F. Foringsforsok paa laks i vinterstidmed ulike forsammensetninger, *Norsk Fiskeoppdrett*, 4 (April), 7, 1982.
376. Viken, N. I. Syrekonservert fisk og fiskeaffall-surfor, *Norsk Fiskeoppdrett*, 10 (October), 4, 1980.
377. Fossheim, E., and G. Parmann. Kontroll med fiskeforet, in: *Norsk Havbruk 1989*, Georg Parmann Presseservice, Nesollhöla, Norway, 54, 1989.
378. Refstie, T. Produksjon av smolt og settefisk, in: T. Gjedrem (Ed.), *Oppdrett av Laks og Aure*, Landbruksforlaget, Oslo, 1979, 96.
379. Isaksson, A. The production of one-year smolts and prospects of producing zero-smolts of Atlantic salmon in Iceland, using geothermal resources, *Aquaculture*, 45, 305, 1985.
380. Hendrix, M. A. Culture techniques for Atlantic salmon (*Salmo salar*) in the Northwestern United States, paper delivered at the 2nd USA-USSR Sym. Reproduction Rearing, and Management of Anadromous Fishes, Seattle, February 8—10, 1990, 1, unpublished.
381. Kittelsen, A. Drift av Klekkeri, in: T. Gjedrem (Ed.), *Oppdrett av laks og aure*, Landbruksforlaget, Oslo, 1979, 87.
382. Ingebrigtsen, O., and O. Torrisen. Utstyr, Metoder og tabeller, Vannet, In: O. Ingebrigtsen (Ed.), *Akvakultur, Oppdrett av Aksefisk*, NKS- forlaget, Oslo, 278, 1982.
383. Needham, T. Salmon smolt production, In: L. Lindsey and T. Needham (Eds.), *Salmon and Trout Farming*, Ellis Horwood series in Aquaculture and Fisheries Support, Chichester, 1988, 86.
384. Ingebrigtsen, O. Fra Yngel til ferdig settefisk, in: O. Ingebrigtsen (Ed.), *Akvakultur, Oppdrett av laksefisk*, NKS-forlaget, Oslo, 1982, 104.

385. Eriksson, T. Sea releases of Baltic salmon; increased survival using a delayed release technique, in: *Proc. Fisheries Bioengineering,* Symp., American Fisheries Society, Bethesda, MD, in press.
386. Gudjonsson, T. Smolt rearing techniques, stocking and tagged adult salmon recapture in Iceland. International Atlantic Salmon Foundation, New York, Spec.Publ.Ser. 4, 227, 1973.
387. Isaksson, A. Salmon ranching in Iceland, in: J.E.Thorpe (Ed.), *Salmon ranching,* Academic Press, London, 1980, 131.
388. Hansen, T. Produksjon av "normalsmolt" og "halvtaarssmolt," *Norsk fiskeoppdrett,* 9 (September), 1986, 1.
389. Isaksson, A., and S. Oskarsson. Returns of comparable microtagged Atlantic salmon (*Salmo salar*) of Kollafjordur stock to three salmon ranching facilities, Report of Institute Freshwater Research, Drottningholm, Sweden, 63, 58, 1986.
390. Brannon, E., C. Feldmann, and L. Donaldson L.. University of Washington zero-age coho salmon smolt production, *Aquaculture,* 28, 195, 1982.
391. Donaldson, L. R., and T. Joyner. The salmonid fishes as a natural livestock, *Scientific American,* 249(1), 51, 1983.
392. Severson, R. F. R. Chitwood, and F. Ratti. Pilot Study of Rearing Atlantic Salmon (*Salmo salar*) in an Accelerated Regime to Under-yearling Smoltification at Oregon Aqua-Foods. Springfield hatchery, Oregon Aqua-Foods (unpublished), 1984.
393. Möller, D. *Norwegian Salmon Farming,* Spec. Publ. Ser. 4, International Atlantic Salmon Foundation, New York, 259, 1973.
394. Needham, T. Atlantics — best in a crisis, *Fish Farmer,* Sept/Oct 1989, 49.
395. Skuladottir, G. U., H. B. Schiöth, E. Gudmundsdottir, B. Richards, F. Gardarson, and L. Jonsson. Fatty acid composition of muscle, heart and liver lipids in Atlantic salmon, *Salmo salar,* at extremely low environmental temperature, *Aquaculture,* 84, 71, 1990.
396. Wedum, K. Driftserfaringer fra landbaserte anlegg, *Norsk Fiskeoppdrett,* 4 (April), 50, 1988.
397. Gjedrem, T., Avlesarbeidet, in: T. Gjedrem (Ed.), *Oppdrett Av Laks og Aure,* Landbruksforlaget, Oslo, 173, 1979.
398. Naevdal, G. Kjönsmodning I, *Norsk Fiskeoppdrett,* 6 (June), 16, 1986.
399. Gunnes K. Produksjon av matfisk i sjö, in: T. Gjedrem (Ed.), *Oppdrett av Laks og Aure,* Landbruksforlaget, Oslo, 1979, 133.
400. Storebakken, T. För og Foring ved lave sjötemperaturer, *Norsk Fiskeoppdrett* 1 (January), 10, 1982.
401. Saunders, R. L., E. B. Henderson, and P. R. Harmon. Effects of photoperiod on juvenile growth and smolting of Atlantic salmon and subsequent survival and growth in sea cages, *Aquaculture,* 45, 55, 1985.
402. Langdon, J. S. Smoltification physiology in the culture of salmonids, in: J. F. Muir and R. J. Roberts (Eds.), *Recent Advances in Aquaculture,* Vol. 2, Croom Helm, London, 79, 1985.
403. Hansen, L. P., B. Jonsson, R. I. G. Morgan, and J. E. Thorpe. Influence of parr maturity on emigration of smolting Atlantic salmon (*Salmo salar*). *Can. J. Fish. Aquat. Sci.,* 46, 410, 1989.
404. Lincoln, R. Putting the pressure on triploidy, *Fish Farmer,* Sept/Oct 1989, 50.
405. Egidius, E. Sjukdom og hygiene i fiskeoppdrett, in: O. Ingebrigtsen (Ed.), *Akvakultur, Oppdrett av Laksefisk,* NKS-Forlaget, Oslo, 1982, 214.
406. Post, G. W. *Textbook of Fish Health,* TFH Publications, Jersey City, New Jersey, 1983, 1.
407. Eggset, G. FURUNKULOSE-en frygtet fiskesygdom, *Norsk Fiskeoppdrett,* 3 (March), 1989, 46.
408. Raa, J. Sykdom, in: E. Fossheim and G. Parmann (Eds.), *Norsk Havbruk,* G. Parmann Presseservice, Nesoddhögda, Norway, 1989, 64.
409. Haastein, T., and S. O. Roald Sjukdommer og sjukdomsbekjempelse, in: T. Gjedrem (Ed.), *Oppdrett av laks og aure,* Landbruksforlaget Oslo, 1979, 250.
410. Haastein, T. 1988. Infeksios hematopoetisk nekrose (IHN)-En ny sjukdom europeisk fiskeoppdrett, *Norsk Fiskeoppdrett,* 4 (April), 35, 1988.
411. Hansen, L. P., and T. A. Bakke. Flukes, genetics and escapees, *Atlantic Salmon Journal,* Autumn, 1989.
412. Thorud, K., T. Lunder, T. T. Poppe, R. A. Holt, and J. Rohovec J. Ny sykdom paavist paa norsk settefiskanlegg, *Norsk Fiskeoppdrett,* 10 (October), 44, 1989.
413. Edwards, D. Vann og vannkvalitet, in: T. Gjedrem (Ed.), *Oppdrett av Laks og Aure,* Landbruksforlaget, Oslo, 1979, 39.
414. Tangen, K. Nytt tilfelle av fiskedöd foraarsaket av planktonalger, *Norsk fiskeoppdrett,* 4 (April), 5, 1982.
415. Fossheim, E. and G. Parmann. Gullaaret, in: E. Fossheim and G. Parmann (Eds.), *Norsk Havbruk,* Parmann Presseservice, 1989, 13.
416. Moltu, T. SFT, fiskeoppdrett og problemomraader mot aar 2000, *Havbruk,* 6, 21, 1989.
417. Hemmingsen, A. R., R. A. Holt, R. D. Ewing, and J. D. McIntyre. Susceptibility of progeny from crosses among three stocks of coho salmon to infection by *Ceratomixa shasta, Trans. Am. Fish. Soc.,* 115, 492, 1986.

418. Mellergaard, S. AAlens svommeblaereorm, *Anguillicola*-En ny parasit i den europaeiske aalebestand, *Nordisk Aquakultur,* 2, 50, 1988.
419. Lichatowich, J. A., and J. D. McIntyre. Use of hatcheries in the management of Pacific anadromous salmonids, in: Dadswell, U. J., R. J. Klanda, C. U. Moffit, R. L. Saunders, R. A. Rulifson, and J. E. Cooper (Eds.), *Common Strategies of Anadromous and Catadromous Fishes,* American Fisheries Society, Bethesda, Maryland, 1987, 131.
420. Mayo, R. D. 1988. An Assessment of Private Salmon Ranching in Oregon, a report to the Oregon Department of Fish and Wildlife, unpublished, 1988, 1.
421. Hindar, K., N. Ryman, and F. Utter. Genetic effects of cultured fish on natural fish populations, *Can. J. Fish. Aquat. Sci.* 48, 945, 1991.
422. Standal, N. Sikring av gener-genforurensning, *Norsk Fiskeoppdrett,* 1 (January), 88, 1988.
423. Bentsen, H. B. 1989. Genetisk paavirkning av villaks fra römt oppdretts-laks, A report to the Norwegian committee of biotechnology, Ås, Norway, unpublished, 1989, 1.
424. Youngson, A. F. Can the factors likely to limit the crossing of wild fish and fish of farmed origin be identified?, Abstract, joint NASCO/ICES meeting on genetic threats, Dublin, Ireland May 23, 1, 1989.
425. Hanssen, O. Dumping av lakserogn og smolt?, *Norsk Fiskeoppdrett,* 2 (February), 1, 1987.
426. OECD. Aquaculture, Developing a New Industry, Organization for Economic Development and Cooperation, Paris, 1989, 1.
427. Hanssen, O., B. Gullestad, and G. E. Blalid. Lakseproduksjonen Rundt 160,000 tonn I 1990, *Norsk Fisheoppdrett,* 1 (January), 6, 1991.
428. Hanssen, O. Smoltoverskudd paa 13 millioner i aar, *Norsk Fiskeoppdrett,* 1 (January), 36, 1987.
429. Hanssen, O. 1988. Smolttilbudet i 1989, *Norsk Fiskeoppdrett,* 11 (November), 4, 1988.
430. Aamlid, T. 1981. Stotte til Fiskeoppdrettsnaeringa, *Norsk Fiskeoppdrett,* 5/6 (May/June), 18, 1981.
431. Crutchfield, J. A. 1989, Economic aspects of salmon aquaculture, *Northwest Environmental Journal,* 5, 37, 1989.
432. Hanssen, O. Matfiskkonsesjoner og Volumbegrensning, *Norsk Fiskeoppdrett,* 10 (October), 6, 1980.
433. Nordstrand, L. 1989, Skotsk oppdrettsnering mindre effektiv enn norsk, *Norsk Fiskeoppdrett,* 3 (March), 28, 1989.
434. Anon. Inverness 1988, *Aquaculture Ireland,* 36, 1, 1988.
435. Needham, T. 1985. The path of progress in the Western Isles, *Fish Farmer,* July/August, 27, 1985.
436. Munro, A. L. S. Personal communication, 1990.
437. Anonymous. Scottish scene, an industry in the making, *Fish Farmer,* July/August, 1, 1985.
438. Reinert, A. Faeroyene, Stor expansion innen oppdrett, *Norsk Fiskeoppdrett,* 6 (June), 36, 1985.
439. Reinert, A. Personal communication.
440. Hanssen, O. Faeroyenes Oppdrettspioner satser for fremtida, *Norsk Fiskeoppdrett,* 12 (December), 46, 1989.
441. Haraldsen, M., S. Horsdal, H. Pedersen, K. Rasmussen, and E. Sundstein. *Fiskeopdraet paa Faeröerne,* a report prepared at the Commercial College in Copenhagen, Vejleder Dennis Clausen, 1989, 1.
442. Hanssen, O. Smolt-debatt og dumpningsak paa Faeroenes forste oppdrettsting. *Norsk Fiskeoppdrett,* 2 (February), 19, 1990.
443. Wilkins, N. P. *Ponds, Passes and Parcs in Victorian Ireland,* Glendale Press, Dublin, 1989, 1.
444. Browne, J. Personal communication, 1990.
445. Joyce, J. Association news, *Aquaculture Ireland,* 38, 5, 1989.
446. Doyle, J. Personal communication, 1991.
447. Keilthy, L. Salmon review 1988/1989, *Aquaculture Ireland,* 41, 16, 1989.
448. Commercial Aquaculture in Canada, Communications Directorate, Department of Fisheries and Oceans, Ottawa, 1988, 1.
449. Department of Fisheries and Oceans. Long Term Production Outlook for the Canadian Aquaculture Industry, Economic and Commercial Analysis, Report No. 13, by Price Waterhouse Management Consultants, Vancouver, British Columbia, 1989, 1.
450. Gudjonsson, T. Salmon, trout and char in Iceland, in: *Country and Population,* Central Bank of Iceland, Reykjavik, 1986, 14.
451. Arnfinnsson, J. and V. Johannsson. Aquaculture Production in Iceland 1989, Institute of Freshwater Fisheries Report, Reykjavik, 1989, 1.
452. Isaksson, A. Salmon ranching in Iceland, in: *Wild Salmon — Present and Future,* proceedings of an international conference held at Sherkin Island Marine Station, Ireland, 16—17 September 1988, 96.
453. Moltu, T. Markeder for laks og perspektiver, *Havbruk,* 6, 9, 1989.
454. Hanssen, O. Lakseprisene, *Norsk Fiskeoppdrett,* 6 (June), 2, 1989.
455. Anonymous. Norway cuts harvest, *Fish Farming International,* December, 1989, 1.

456. Aasgaard, B. Minsteprisane ned 4-8 kroner, -ikkje kvotereguleringnu, *Norsk Fiskeoppdrett,* 6 (June), 3, 1989.
457. Nasaka, Y. Salmonid programs and public policy in Japan, in, W. J. McNeil (Ed.), *Biological, Economic and Policy Issues,* Oregon State University Press, Corvallis, Oregon, 1988, 25.
458. Iwata, M. *Smoltification, Seawater Adaptability and the Growth of Coho Salmon (Oncorhynchus kisutch)* in Relation to the Culture Environments in Japan, Proceedings of Aquaculture International Congress, September, British Pavilion Corporation, 1988, 391.
459. Endo, N. On the commercial coho salmon culture, in, Preliminary Report of the Salmonid Workshop on Biological and Economic Optimization of Smolt Production, Versailles Summit Working Group, Tokyo, Japan, 22—25 January, 1985, Ministry of Agriculture, Forestry, and Fisheries, Japan, 1985, 146.
460. Anonymous. Pen-raised chinook, *Bill Atkinson's News Report* p. 202, 1987.
461. Mottet, M. G. Factors Leading to the Success of Japanese Aquaculture, with an Emphasis on Northern Japan, Washington Dept. Fisheries Tech. Rep. 52, 1980, 1.
462. Asada, Y. License limitation regulations: the Japanese system. *J. Fish. Res. Bd. Can.,* 30, 2085, 1973.
463. Nasaka, Y. Trip Report on Japan's Coho Salmon Pen Culture, U.S. Embassy/Japan memorandum, Fisheries Attaché, 12 August 1987, 1.
464. Toole, C. Report of a Study Trip to Japan. II, Salmonid Aquaculture and Fisheries, University of California Sea Grant Extension Program, Working Paper PT-48, 1, 1988.
465. Corey, P. D., D. A. Leith, and M. J. English. A growth model for coho salmon including effects of varying ration allotments and temperature, *Aquaculture,* 30, 125, 1983.
466. Brett, J. R. Temperature tolerance in young Pacific salmon, genus *Oncorhynchus, J. Fish. Res. Bd. Can.,* 9, 265, 1952.
467. Iwata, M., and C. Clarke. Culturing coho Japanese-style. *Can. Aquaculture,* 3, 28, 1987.
468. Harache, Y., G. Boeuf, and P. Lasserre. Osmotic adaptation of coho salmon (*Oncorhynchus kisutch*) Walbaum. III. Survival and growth of juvenile coho salmon transferred to sea water at various times of the year, *Aquaculture,* 19, 253, 1980.
469. Bunya, T. Report on the inquiry into the strategies to combat self-pollution existing in the silver salmon aquaculture industry, Sea Water Adaptation Study. *Miyagi Prefecture Kesenuma Fisheries Experimental Station Report,* 88, 57, 1983 (in Japanese).
470. Yoshida, F. Seawater culture of coho salmon in Japan, in: Preliminary Report of the Salmonid Workshop on Biological and Economic Optimization of Smolt Production, Versailles Summit Working Group, Tokyo, Japan, 22—25 January, 1985, Ministry of Agriculture, Forestry, and Fisheries, Japan, 1985, 137.
471. Kiyama, K. Report on the inquiry into the strategies to combat self pollution existing in the silver salmon aquaculture industry, Rate of conversion of feed, *Miyagi Prefecture Kesenuma Fisheries Experimental Station Report,* 88, 41, 1983 (in Japanese).
472. Bowden, G. B. *Coastal Aquaculture Law and Policy: A Case Study of California,* Westview Press, Boulder, Colorado, 1981, 1.
473. Wildsmith, H. *Aquaculture: The Legal Framework,* Emond-Montgomery Ltd., Toronto, Canada, 1982, 1.
474. *Les Cultures Marines en France et le Droit,* Centre de droit et d'économie de la mer, Centre National pour l'Exploitation des Océans, Brest, France, 1983, 1.
475. White, S. Anticipating the 'blue revolution': the growth of the salmon farming industry and its public policy implications, *Anadromous Fish Law Memo,* Issue 45, April 1988, 1.
476. Robinette, H. R., J. Hynes, N. C. Parker, R. Putz, R. F. Stevens, and R. R. Stickney. Commercial aquaculture, *Fisheries,* 16(l), 18, 1991.
477. Bowen, J. T. A history of fish culture as related to the development of fishery programs, in: Benson, N.G. (Ed.), *A Century of Fisheries in North America,* Spec. Publ. No. 7, American Fisheries Society, Washington, D.C., 1970, 71.
478. Kirk, R., *A History of Marine Fish Culture in Europe and North America,* Fishing News Books Ltd., Farnham, Surrey, England, 1987., 122.
479. Larkin, P. A., Management of Pacific salmon of North America, in: Benson, N. G. (Ed.), *A Century of Fisheries in North America,* Spec. Publ. No. 7, American Fisheries Society, Washington, D.C., 1970, 223.
480. Brannon, E. and G. Klontz. The Idaho aquaculture industry, *Northwest Environmental Journal,* 5, 23, 1989.
481. Cooper, E. L. Management of trout streams, in: Benson, N. G. (Ed.), *A Century of Fisheries in North America,* Spec. Publ. No. 7, American Fisheries Society, Washington, D.C., 1970, 153.
482. Glude, J. B. Aquaculture for the Pacific Northwest: a historical perspective, *Northwest Environmental Journal,* 5,7, 1989.
483. Scarnecchia, D. L. The history and development of Atlantic salmon management in Iceland, *Fisheries,* 14(2), 14, 1989.

484. Mathews, S. B., and H. G. Senn. *Chum Salmon Hatchery Rearing in Japan,* Washington Sea Grant Program Publication WSG-TA-75-3, University of Washington, Seattle, 1975, 1.
485. Nishiyama, T. Japanese and Soviet attitudes toward aquaculture, *Aquaculture Notes,* University of Alaska Sea Grant Program, Fairbanks, April 1977, 1.
486. Khozin, G. *The Biosphere and Politics,* English edition Belitsky, B., (trans), Progress Publishers, Moscow, 1979, 1.
487. Whitelaw, I. M. Structure and administration of the Scottish salmon farming industry, in: *Aquaculture: A Review of Recent Experience,* Organization for Economic Cooperation and Development, Paris, 1989, 206.
488. Stortinget. *Om Havbruk,* Stortingets Meld. No. 65 (1986-1987), Norges Offentlige Utredninger, Universitetsforlaget, Oslo, 1987, 1.
489. Thorpe, J. E. Ocean ranching — general considerations, in: A. E. J. Went (Ed.), *Atlantic Salmon: Its Future,* Fishing News Books Ltd., Farnham, Surrey, England, 1980, 152.
490. McNeil, W. J., Legal aspects of ocean ranching in the Pacific, in: J. E. Thorpe (Ed.), *Salmon Ranching,* Academic Press, London, 1979, 383.
491. Blumm, M. C. Hydropower vs. salmon: the struggle of the Pacific Northwest's anadromous fish resources for a peaceful coexistence with the federal Columbia River power system, *Environmental Law,* 11, 212, 1981.
492. Shaw, S. and J. F. Muir. *Salmon Economics and Marketing,* Croom Helm, London, 1987, 79.
493. Kerns, C. World Salmon Farming: An Overview with Emphasis on Possibilities and Problems in Alaska, Marine Advisory Bulletin ;#26, Fairbanks, Alaska, December 1986, 1.
494. Alaska Finfish Farming Task Force, Report to the Alaska Legislature, Juneau, January 15, 1990, 1.
495. Herrmann, M, B., H. Lin, and R. C. Mittelhammer. *U.S. Salmon Markets: A Survey of Seafood Wholesalers,* Report No. 90-01, Alaska Sea Grant College Program, Fairbanks, 1990, 1.
496. Ackefors, H. and M. Enell. Discharge of nutrients from Swedish fish farming to adjacent sea areas, *Ambio,* 19, 28, 1990.
497. Underdal, B., O. M. Skulberg, E. Dahl, and T. Aune. Disasterous bloom of *Chrysochromulina polylepsis* (Prymnesiophyceae) in Norwegian coastal waters 1988 — mortality in marine biota, *Ambio,* 18, 265, 1989.
498. Solbé, J. Water quality, in: L. Laird, and T. Needham (Eds.), *Salmon and Trout Farming,* Ellis Horwood Ltd., Chichester, 1988, 69.
499 Fiskeridepartementet. *Akvakultur i Norge: Status og Fremtidsutsikter,* Norske Offentlige Utredninger, No. 22, Universitetsforlaget, 1985, 1.
500. Shaw, S. A. The economics of Scottish salmon farming, in: Aquaculture: A Review of Recent Experience, Organization for Economic Cooperation and Development, Paris, 1989, 241.
501. Dunn, T. P., P. A. Leitz, and S. L. Harris. The salmon aquaculture industry in Canada, in: Aquaculture: A Review of Recent Experience, Organization for Economic Cooperation and Development, Paris, 1989, 273.
502 Whitener, B., Tribal issues in aquaculture and environment, *Northwest Environmental Journal,* 5, 111, 1989.
503. Hershberger, W. K. Directed and "inadvertent" genetic selection in salmonid culture: results and implications for the resource and regulatory approaches, in: W. J. McNeil (Ed.), *Salmon Production, Management and Allocation: Biological, Economic and Policy Issues,* Oregon State University Press, Corvallis, Oregon, 1988, 177.
504. Nordic Council. Policy and technology enable discharges to be reduced, *Northern Europe's Seas: Northern Europe's Environment* (trans.), Norstedts Tryckeri AB, Stockholm, 1989, 170.
505. Haines, T.A. Effects of acid rain on Atlantic salmon rivers and restoration efforts in the United States, in: L. Sochasky (Ed.), *Acid Rain and the Atlantic Salmon,* International Atlantic Salmon Foundation, Spec. Publ. Ser. No. 10, Print'N Press, St. Stephen, New Brunswick, 1981, 57.
506. Sochasky, L., (Ed.), *Acid Rain and the Atlantic Salmon,* International Atlantic Salmon Foundation, Spec. Publ. Ser. No. 10, Print'N Press, St. Stephen, New Brunswick, 1981, 1.
507. Miller, K., and D. L. Fluharty, El Niño and variability in the Northeast Pacific salmon fishery: implications for coping with climate change, in: M. Glantz (Ed.), Global Climate Change, Ocean Variability and Fisheries, Cambridge University Press, Cambridge, in press.
508. Maitland, P. S. The potential impact of fish culture on wild stocks of Atlantic salmon in Scotland, in: D. Jenkins and W. M. Shearer (Eds.), *The Status of the Atlantic Salmon in Scotland,* The Cambrian News, Aberystwyth, Great Britain, 1986, 73.
509. Nehlsen, W., J. E. Williams, and J. A. Lichatowich. Pacific Salmon at the Crossroads: Stocks at Risk from California, Oregon, Idaho and Washington, unpublished manuscript. 1990, 1.
510. Buck, E. H., A. Abel, M. L. Kessler, and E. B. Bazan. Pacific Salmon and Steelhead: Potential Impacts of Endangered Species Act Listings, Congressional Research Service Report for Congress 90-533 ENR, Library of Congress, Washington, D.C., November 16, 1990, 1.

INDEX

A

B

C

D

E

F

P

T

U

V

W

Y

Z